Cretaceous Sea Level Rise

Cretaceous Sea Level Rise

Down Memory Lane and the Road Ahead

Mu Ramkumar

ELSEVIER
AMSTERDAM • BOSTON • HEIDELBERG • LONDON
NEW YORK • OXFORD • PARIS • SAN DIEGO
SAN FRANCISCO • SINGAPORE • SYDNEY • TOKYO

Elsevier
Radarweg 29, PO Box 211, 1000 AE Amsterdam, Netherlands
The Boulevard, Langford Lane, Kidlington, Oxford OX5 1GB, UK
225 Wyman Street, Waltham, MA 02451, USA

ISBN: 978-0-12-805414-7

British Library Cataloguing-in-Publication Data
A catalogue record for this book is available from the British Library

Library of Congress Cataloging-in-Publication Data
A catalog record for this book is available from the Library of Congress

For information on all Elsevier publications
visit our website at http://store.elsevier.com/

CONTENTS

PREFACE

The recent enhanced awareness on the of sea level rise as a response to natural and anthropogenic causes has emphasized the importance of understanding sea level fluctuations of the past especially that of the Cretaceous Period, a time known for extended duration of greenhouse conditions. It has also led to a specific project—IGCP 609 with mandates to address correlation, causes (regional/tectonic or global/eustatic) and consequences of significant short-term (Ka to 100s of Ka) sea level fluctuations in the climatic cycles, the feedback mechanisms, relationship between the sea level high-ocean anoxia and sea level low-ocean oxidation events and to evaluate the evidence for ephemeral glacial episodes or other climate events. Though research has been and is being conducted in these fields and a major thrust has been given since the introduction of sea level cycles and sequence development concepts in the 1970s, the available information is scattered. This book reviews the available information and presents them in terms of current understanding, the existing consensus and conflicts on the causes and consequences of sea level fluctuations during the Cretaceous, and the road ahead. It is more of a stock-taking effort, to inform the readers at advanced undergraduates, post graduates, doctoral and other beginner level researchers, so that, they get to know the developmental history, tools, mechanisms, causes and current understanding on this fascinating field of research and get themselves involved in filling the gaps and improve the scientific understanding.

Our current understanding on the Cretaceous sea level cycles is far from complete. Though sea level highs and lows are recorded in widely separated basins and are linked to the eustasy, consensus on the causative mechanisms is yet to be evolved, principally due to the fact that despite the sea level fluctuations during the Cretaceous were cyclic, they were only periodic. Added to this are the problems of poorly constrained strata, gaps in sedimentary records, lack of appreciation in utilizing multiproxies, complexities introduced by local and regional tectonic and other processes within the eustatic cycles and the lack of consensus on the existence and role of glacial feedback mechanism during the greenhouse conditions. These need to be addressed by the researchers especially through application of multiproxy studies on extended stratigraphic records.

Exhaustive and excellent list of case studies demonstrating multitudes of tools, techniques and methods on wide variety of tectonic, depositional, climatic, and lithological terrains is available that documented a range of sea level cycles and perturbations and attempted explaining them. However, the perturbations among the cyclic nature of Cretaceous sea level cycles have been the focus of many studies, yet consensus on the causes and consequences elude the researchers. Notable among the perturbations are those linked with oceanic anoxia during early part of Upper Cretaceous and those linked with the dwindling of species diversity/population, environmental stress, and extinction during the latest Cretaceous. Though the previous studies recognized suitable stratigraphic sections representing marine and nonmarine realms of these time slices, concerted efforts have to be taken to characterize and create better understanding on the spatiotemporal and causative mechanisms of the cycles and perturbations through multiproxy studies.

Mu Ramkumar

ACKNOWLEDGMENTS

Understanding on this fascinating field of research was aided by the articles cited in this book that were the outcomes of many individuals and research teams who have toiled hard since many decades. Author places on record his deep sense of regard for all these authors. However, author alone is responsible for the views and statements presented. Interest to review and take stock of the current understanding on the Cretaceous sea level fluctuations has been driven by author's nomination to the National Working Group of IGCP–609 by the Geological Survey of India.

Dr. Condice Janco, Publisher, Earth and Environmental Sciences, Elsevier Science and Technology Books, Dr. Louisa Hutchins, Acquisitions Editor, and other members of the editorial team of Elsevier, the anonymous reviewers who have recommended my proposal and those at the helm of administrative and technical approval committees at Elsevier are thanked profusely for their encouragement and continued support that helped hone my skills and scientific acumen.

My involvement on this subject and scientific collaboration with national and international academic and research institutions has been and is being supported by research grants from various organizations, namely, the Alexander von Humboldt Foundation, Germany, the University Grants Commission, the Council of Scientific and Industrial Research, the Department of Science and Technology, the Oil Industry Development Board, the Oil and Natural Gas Corporation Limited, India, and the German Research Foundation, Germany, etc., for which I am thankful.

Many individuals, including, but not limited to, Prof. Vladan Radulovic, Faculty of mining and Geology, Belgrade University, Prof. David Menier Jean Claude, University of South Brittany, France, Prof. Dmrity Ruban, Southern Federal University, Russia, Dr. Jyotsana Rai, Birbal Sahni Institute of Palaeobotany, Lucknow, India, Dr. Zsolt Berner, Institute of Mineralogy and Geochemistry, Karlsruhe Institute of Technology, Germany, and Prof. Franz T.

Fürsich, GeoZentrum Nordbayern, Fachgruppe PaläoUmwelt, Friedrich-Alexander-Universität Erlangen-Nürnberg, Germany, are thanked for encouragement and academic support. The team at Elsevier's production unit including Mohanapriyan Rajendran had done excellent work and thanked for professional and flawless handling.

My wife A. Shanthy, daughter, Ra. Krushnakeerthana, and son Ra. Shreelakshminarasimhan are thanked for their support, understanding, and also for forbearing my absence while authoring this book.

Above all, I submit my thankfulness unto the lotus feet of The Lord Shree Ranganayagi samedha Shree Ranganatha, for his boundless mercy showered on me, and by whose ordinance those mentioned above have shouldered the responsibilities either actively or passively and helped me to complete this work.

Mu Ramkumar

CHAPTER 1

Introduction

The modern sea level rise, which commenced 21,000 years ago, may continue for about another 40,000 years, should the natural process is allowed to continue, under the influence of shape of Earth's orbit. However, the advent of modern human species, and the technological prowess the species learned to unleash to the detriment of the earth processes have interfered significantly to change the cyclic process of global ice volume changes and attendant environmental and climatic conditions. The introductory note of the *UNESCO IGCP-609 Climate-environmental deteriorations during greenhouse phases: Causes and consequences of short-term Cretaceous sea level changes*, states that *"the recent rise in sea level in response to increasing levels of atmospheric greenhouse gases and the associated global warming is a primary concern for the society. Evidence from the Earth's history indicates that the glacial-interglacial and other ancient sea level changes occurred at rates an order of magnitude or higher than that observed at present. To predict future sea levels, we need a better understanding of the record of past sea level change"*. The Late Cretaceous has been a target of intensive study for past climatic and sea level fluctuations, due to excellent preservation of both oceanic and epicontinental strata, development of detailed biostratigraphic, geochronologic, and sequence stratigraphic frameworks, and a series of global scale biogeochemical perturbations that record major disruptions in the Earth's carbon cycle and climate (Joo and Sageman, 2014). The interaction between sea level and biodiversity is important on long-term as well as short-term time scales (Holland, 2012). Similarly, the majority of the most profound extinctions in the Earth's history are in one way or another related to sea level changes (Hallam and Wignall, 1999). These changes are often abrupt and possibly related to the loss of ecospace (regressions) or anoxia (transgression). For all these reasons, accurate knowledge on the sea level is essential (Munnecke et al., 2010).

Cognizant on the control exercised by the eustatic sea level fluctuations over the stratal patterns and the facies distribution, the concept

Cretaceous Sea Level Rise. DOI: http://dx.doi.org/10.1016/B978-0-12-805414-7.00001-1

of global sea level changes was introduced. It provided an impetus for correlating stratigraphic records with their counterparts located elsewhere. The relative sea level cycles (Figure. 1.1), first published by Vail et al. (1977) and revised by Haq et al. (1987, 1988), espoused that the sedimentary sequences are produced principally under the influence of sea level cycles, durations of which vary between few tens of millions of years (first-order cycle) and few million years (third-order cycle). Successive studies have shown that distinct sedimentary sequences (though the use of the term "sequence" itself was found to be diverging and misleading–van Loon, 2000) could be traced to sea

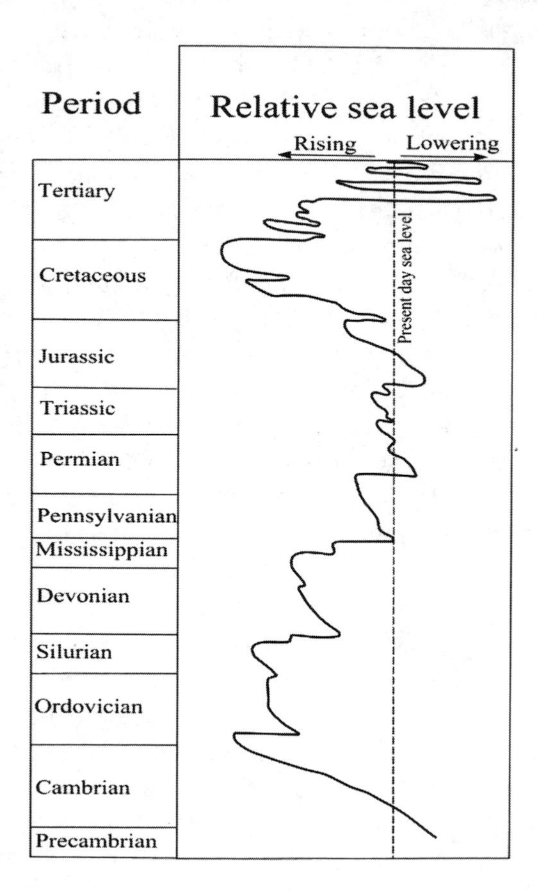

Figure 1.1 ***Eustatic sea level curve which came to be known as Vail/Exxon curve.*** *The morphology, causes of sudden sea level changes and the advocacy of using this curve for chronostratigraphic determination of strata were severely criticized by the scientific community, though, the concepts of shoreline shift as a result of tectonics, relative sea level fluctuation, climate and sediment flux were accepted, albeit reluctantly and with skeptism. See Miall (1991) and Miall and Miall (2001) for a detailed review on this debate. See Miall (2009), Catuneanu (2006), and Catuneanu et al. (2011) for current understanding and prevailing consensus. Source:* Modified from Vail et al. (1977).

level cycles of up to infra seventh order (Nelson et al., 1985; Williams et al., 1988; Carter et al., 1991). Vail et al. (1977) stated that the sea level chart published by them is incomplete and cycles of varying order could be added as the studies on sedimentary sequences progress; so that, more complete chart could be constructed. The aim behind this statement was to incorporate sea level cycles at the Milankovitch scale, to which the response of the sedimentation system is proved beyond reasonable doubt (Carter et al., 1991). Hays et al. (1976) convincingly demonstrated that the climatic records were dominated by frequencies characteristic of variations in the Earth's tilt, precession, and eccentricity relative to the Sun. In the years since, numerous studies have upheld the validity of the Milankovitch climatic cycles in terms of 100, 41, and 23 Ka orbital periods that influence or control variations in sea-floor spreading (Quilty, 1980), global ice volume, thermohaline circulation, continental aridity and run off, sea surface temperature, deep ocean carbonate preservation, and atmospheric CO_2 and methane concentrations (Raymo et al., 1997). Cyclic sedimentation has been documented in numerous sedimentary basins and there are many lines of evidences that relate those cycles to short-term (Milankovitch band) glacioeustatic pulses (Laferriere et al., 1987; Grammer et al., 1996; Gale et al., 2002). Park and Oglesby (1991) are of the opinion that the Cretaceous deposits are characterized by cyclic deposition of carbonates with periodicities similar to the Milankovitch band. While examining the compiled data on the sediment volumes (mass) or sediment fluxes of the continental and marine subsystems to determine the complete routing in terms of mass conservation for specific time periods since the Late Glacial Maximum as well as the Cenozoic, Hinderer (2012) reported that the response times of the large sedimentary systems are within the Milankovitch band. Hilgen et al. (2014) opined that despite fragmentary sedimentation, stratigraphic continuity as revealed by cyclostratigraphy unequivocally supports the dominant role of depositional processes at the Milankovitch scale. On the contrary, Immenhauser (2005) stated that it is generally assumed that the effects of tectonism, sediment accumulation and compaction, and eustasy on the accommodation change cannot be untangled on a regional scale. In the time domain of few Ma and less, global cycle correlation is often unreliable and the effects of isostasy on the sea level show strong provincial variations. In contrast, on a regional scale, cycle correlation is more precise and regional tectonism has predictable limitations.

The relative sea level chart based on the seismic stratigraphic studies at Exxon, first presented by Vail et al. (1977) and later revised by Haq et al. (1987, 1988), recognized about 100 global sea level changes. Mitchum and Van Wagoner (1991) reported that a hierarchy of interpreted eustatic cyclicity in siliciclastic sedimentary rocks has a pattern of superposed cycles with frequencies in the ranges of 9–10, 1–2, 0.1–0.2, and 0.01–0.02 Ma (second- through fifth-order cyclicity, respectively). While examining the $\delta^{18}O$ of Phanerozoic seawater, Veizer et al. (1997) observed the presence of high-frequency cycles within first-order cycle. Strauss (1997) recorded fourth-order cycles of sulfur isotope that stacked up to form third-order cycle fluctuations that in turn were accommodated within second-order cycles. Goldhammer et al. (1991) showed that the sequences of Paradox Basin exhibited a distinct cyclicity characterized by a hierarchical stacking pattern such that fifth-order shallowing upward cycles group into fourth-order cycles, which in turn stack vertically into part of a third-order cycle. McGhee (1992) and Gjelberg and Steel (1995) observed high-frequency sea level changes within a major sea level cycle. A detailed study of the Great Limestone Cyclothem of the northern Pennines (Alston Block) by Tucker et al. (2009) revealed that within the transgressive carbonates, the beds, averaging 75 cm in thickness and defined by millimeter-shale partings or centimeter-mudrock layers, form two thinning-upward to thickening-upward bed-sets. Oxygen isotope and strontium trace element data also revealed patterns of increasing and decreasing values through the limestone, which broadly correspond to the bed-thickness cycles. The beds are interpreted as millennial-scale cycles, the result of high-frequency, arid–humid climatic fluctuations. The bed-sets are interpreted as the response to a longer term arid–humid climate and sea level cycle, driven by the precession rhythm. Strasser et al. (1999) studied the Oxfordian and Berriasian sections representing shallow-water, carbonate-dominated sedimentary systems in the Swiss and French Jura, in Spain, and in Normandy. They all displayed a hierarchical stacking of depositional sequences. The superposition of high-frequency sea level changes on a long-term sea level trend led to repetition of diagnostic surfaces, defining sequence-boundary and maximum-flooding zones wherein the corresponding high-frequency surfaces are well developed. Chronostratigraphic tie points permitted estimation of the duration of large-scale sequences. This time control and the observed hierarchical stacking suggested that the high-frequency sea level changes were controlled by climatic cycles in the Milankovitch frequency band.

The stacking pattern of the cycles of sea level changes has not been questioned though consensus eludes on the causes and magnitudes of sea level variation (eg, Miller et al., 2003, 2004), absolute timings of these changes, and the utility of sea level charts for chronostratigraphy (Summerhayes, 1986; Hancock, 1993). Lack of consensus is such that, there is no established/accepted sea level chart across the K/Pg (Esmeray-Senlet et al., 2015), a boundary section that has been most intensively studied by the geoscientists! A much larger pattern (about 200 Ma) is interpreted as tectonically controlled eustasy probably related to the sea-floor spreading rates. Heller and Angevine (1985) suggested formation of Atlantic-type oceans with its attendant changes in the area/age distribution of the global ocean floor and the formation of extensional passive margins might be the primary reason for this first-order sea level changes and the role of oceanic floor spreading rate might be secondary. The sedimentary records of the Earth are the results of past tectonic and climatic events (Zhang et al., 2001; Armitage et al., 2011). As the sea level fluctuations are intimately coupled with climatic conditions and are often overprinted by tectonic evets, an ability/tool/method to accurately differentiate them is necessary.

From this review, it emerges that the history of sea level fluctuations was under the influences of myriad varieties of causes and varied durations. Though pragmatic estimates on the periodicities could be made, reliable and comparable stratigraphic records for authenticating such estimates are scarce or are limited to few time slices. As explained in following chapters, these well-constrained time slices and their stratigraphic records, despite being intensively studied, lack consensus on their sea level trends. With this background, in this book, an attempt is made to collate the available information on sea level fluctuations, causes, and the current understanding on the Cretaceous sea level cycles through extensive review of the classic and recent literature. However, it is not the task of this book to present an inventory of publications on Cretaceous sea level cycles; instead, attempt is made to take stock of our current understanding as explicit in the available published information and case studies, find the gaps and conflicts, and present them for future research to fill the lacuna and evolve a consensus.

CHAPTER 2

Trends, Timings, and Magnitudes

The sea level curve of Haq et al. (1988) and the long-term trend in its updated version (Haq, 2014; Cloeting and Haq, 2015) depict an increase in the frequency of sea level oscillations since the Mid-Cretaceous under the influence of astronomic factors with a periodicity of 100,000 years superimposed on a gradual rise (Figure 2.1). In the frame of the general Early Cretaceous sea level rise, the Late Barremian–Early Aptian represents an interval of generally low sea level, with a major rise in sea level beginning in the Late Aptian. This sea level change is superimposed by several second- and third-order transgressive–regressive facies and sea level cycles (Bachmann and Hirsch, 2006). Hancock and Kauffman (1979) estimated that during the Late Cretaceous (94–64 Ma) sea level was up to 650 m higher than at the start of the Albian, when it was perhaps about the same as it is today. The worldwide recognizable peak transgressions of the Late Albian, Early Turonian, Early Coniacian, Middle Santonian, and Late Campanian involved rises of sea level at rates of about 10–90 m/Ma. Ruban et al. (2010) have demonstrated that the sea level rise during Cretaceous was not the biggest among other Phanerozoic rises. In addition, there are uncertainties associated with the Cretaceous sea level cyclicity (Ruban, 2015). The peak regressions are more difficult to measure, but were usually faster and involved falls of sea level of about 95–170 m/Ma. A range of sea level fluctuation between 25 and 120 m was suggested (Boulila et al., 2011) to be possible by glacioeustatic cause. There have been eight major marine transgressions on the western margin of Australia since the Late Santonian. These are equated with sea level highs which are believed to be worldwide (Quilty, 1980). Li et al. (2000) recognized sea level regressions during Late Campanian (74.2, 73.4–72.5, and 72.2–71.7 Ma), Early Maastrichtian (70.7–70.3, 69.6–69.3, and 68.9–68.3 Ma), and Late Maastrichtian (65.45–65.3 Ma) out of which, four were found to be global eustatic in nature (74.2, 70.7–70.3, 68.9–68.3, and 65.45–65.3 Ma).

Cretaceous Sea Level Rise. DOI: http://dx.doi.org/10.1016/B978-0-12-805414-7.00002-3

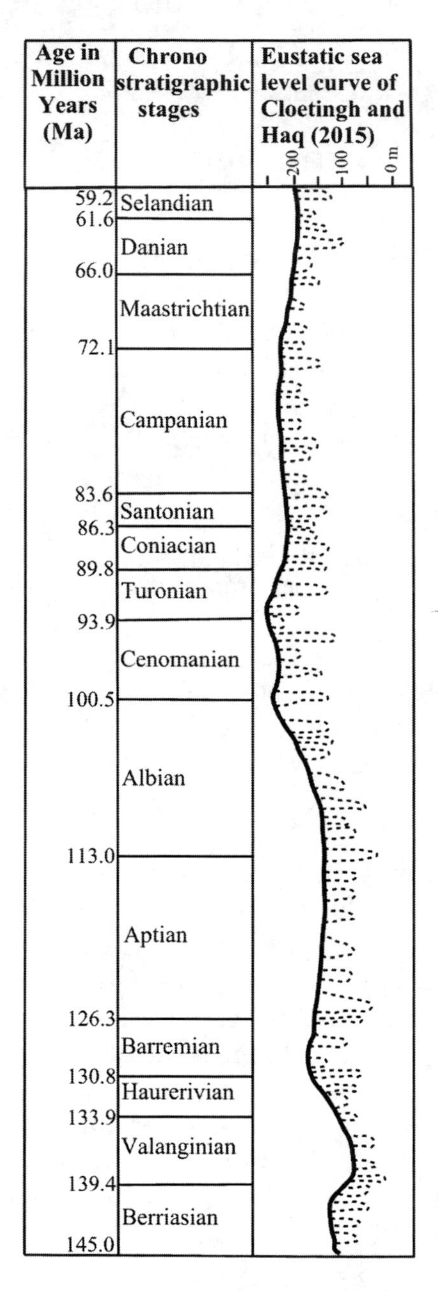

Figure 2.1 ***Long-term and high-frequency sea level fluctuations during Cretaceous.*** *Note that this recently published curve differs significantly from the initially published Vail et al. (1977), and Haq et al. (1987, 1988), essentially in terms of change from a sudden sea level falls to a sinuous morphology, signifying the gradual change rather than sudden switches between rise and fall. The curve depicts the long-term trend (solid line) within which about 58 short-term cycles (dashed line) are embedded.* Modified from Cloetingh and Haq (2015).

Miller et al. (2005) reviewed the Phanerozoic sea level record and presented a new curve for the past 100 million years. They also stated that long-term sea level peaked at 100 ± 50 m during the Cretaceous, implying that ocean-crust production rates were much lower than previously inferred. Immenhauser and Scott (1999) found regional variability within seven Albian sea level curves from localities around the globe (with durations of 0.3–2.7 Ma) and concluded that their global correlation cannot be demonstrated with the available stratigraphic resolution. Frequency trends show that the Albian sea level was statistically stable from the Late Early to the Early Late Albian, whereas it fluctuated rapidly during the earliest and latest Albian. Groetsch (1996) recognized 252 peritidal cycles in the Barremian-Albian Gavrovo-Tripolitza Zone (NW Greece) and correlated with coeval deposits located in Mexico. However, he has cautioned against direct correlation of the strata due to the time gaps of different lengths in the succession during sea level lowstands may bias the direct translation of cycle stacking pattern into third-order sea level curves.

Recently revised and updated Cretaceous sea level curve (Haq, 2014) shows that average sea levels throughout the Cretaceous remained higher than the present day mean sea level (PDMSL) (75–250 m above PDMSL). Sea level reached a trough in Mid Valanginian (~75 m above PDMSL), followed by two high points, the first in Early Barremian (~160–170 m above PDMSL) and the second, the highest peak of the Cretaceous, in earliest Turonian (~240–250 m above PDMSL). The curve also displays two ~20 Ma-long periods of relatively high and stable sea levels (Aptian through Early Albian and Coniacian through Campanian). The short-term curve presented by Haq (2014) identifies 58 third-order eustatic events in the Cretaceous, most have been documented in several basins, while a smaller number are included provisionally as eustatic. The amplitude of sea level falls varies from a minimum of ~20 m to a maximum of ~100 m and the duration varies between 0.5 and 3 Ma.

These information, culled from many different publications based on varied geographic locations, depositional settings, and proxies, suggest that the Cretaceous Period had experienced significant sea level fluctuations in a cyclic fashion of varying durations which in turn are recognizable from stratigraphic sections located at geographically widely separated regions (meaning eustatic nature of the sea level fluctuations). However, imprinted upon these are local and regional fluctuations and also certain anomalous fluctuations. The periodicities and the magnitudes of these fluctuations require varied explanations on the causative mechanisms.

CHAPTER 3

Causes and Mechanisms

3.1 ENDOGENIC PROCESSES AND CAUSES

As it lies at the intersection of Earth's solid, liquid, and gaseous components, the sea level links the dynamics of the fluid part of the planet with those of the solid part of the planet (Conrad, 2013) and hence, any attempt to understand the sea level fluctuations can never be satisfactory if studied in isolation. Based on a review of sea level research conducted during past few decades, Conrad (2013) demonstrated that the solid components of the Earth exert a controlling influence on the amplitudes and patterns of sea level change across time scales ranging from years to billions of years. According to him, on the shortest time scales (10^0-10^2 year), the elastic deformation causes the ground surface to uplift instantaneously near deglaciating areas while the sea surface depresses due to diminished gravitational attraction. On the time scales of 10^3-10^5 year, the solid Earth's time-dependent viscous response to deglaciation also produces spatially varying patterns of relative sea level change, with centimeters-per-year amplitude, that depend on the time history of deglaciation. These variations, on average, cause net sea-floor subsidence and therefore global sea level drop. On time scales of 10^6-10^8 year, convection of Earth's mantle also supports long-wavelength topographic relief that changes as continents migrate and mantle flow patterns evolve. This changing "dynamic topography" causes meters per millions of years of relative sea level change, even along seemingly "stable" continental margins, which affected all stratigraphic records of Phanerozoic sea level. In a lucid description of this "glacio-isostacy," Engelhart et al. (2011) stated that growth of glacial ice volume results in subsidence of land beneath the ice mass, which is compensated for by an outward flow of mantle material that uplifts a peripheral bulge around the ice margin. During thawing of glacial ice this loading is diminished, resulting in uplift of land beneath the melted.

Prevalent large-scale latitudinal eustatic sea level changes, besides normal tectono-eustasy were reported by Mörner (1981) based on the

Cretaceous Sea Level Rise. DOI: http://dx.doi.org/10.1016/B978-0-12-805414-7.00003-5

analysis of the geographical distribution of sea level changes throughout Cretaceous time. These represent south-to-north migrating geoidal-eustatic highs and lows, or gravitational "drop motions." It implies that global eustasy is an illusion and that regional eustasy should be the new key phrase. Van Sickel et al. (2004) stated that determination of the record of global sea level change (eustasy) is one of the more complex problems. Any estimate derived from continental or continental margin data is necessarily a combination of both eustatic and tectonic changes, and is also impacted by local or regional sedimentary processes and their climatic controls.

Cogné et al. (2006) modeled age versus surface distribution of oceanic lithosphere, the effects of oceanic plateau formation and ice cap development to examine the sea level pattern since 180 Ma and found that the predicted sea level curve resembled exactly the first-order sea level variations. These authors also reported initial slower phase of sea level rise (by 90–110 m), during 120 and 50 Ma, followed by a faster phase of sea level drop. Pitman (1978) proposed that the Cretaceous high sea level stand (~125 Ma) and the subsequent lowstand (95 Ma) could be explained solely by expansion and contraction of the mid-oceanic ridge system. According to Wyatt (1986), variation in sea level for most of the Mesozoic, requiring explanation by mechanisms other than the area–elevation effect, was of the order of 160 m. Wyatt (1986) also stated that a major lowering of sea level started in the Late Cretaceous and continued into the Cenozoic. This major change can be explained by changes in ocean ridge volumes. High rates of oceanic crust production during the Cretaceous have been causally linked to high global sea level and global CO_2 due to increased outgassing.

Marzocchi et al. (1992) observed a highly negative correlation (significance level <0.01) with long-term sea level variations that follow by about 6–9 Ma and instability in the rate of occurrence of geomagnetic reversals. These authors interpreted the phenomena as a link between crustal and deep interior Earth processes with that of the mean sea level. The study of Spasojevic and Gurnis (2012) employed a hybrid approach combining inverse and forward models of mantle convection and accounted for the principal contributors to long-term sea level change, distribution of ocean floor age, dynamic topography in oceanic and continental regions, and the geoid. They inferred the relative importance of dynamic versus other factors of sea level change, determined the time-dependent patterns of dynamic subsidence and uplift of continents, and derived a sea level curve and found that both dynamic factors and the evolving distribution of sea-floor age are important in controlling sea level.

On the basis of review of nature of repetitions and cycles in stratigraphic records and their formative causes, Schwarzacher (2000) suggested that the earth's crust has reached a state of self-organized criticality. As a consequence, the sequence formation might be part of the complexity and an attribute of the earth's crust and history. Temporal variation of mantle warming by supercontinental insulation was suggested to have a nontrivial consequence for the long-term eustasy (Korenaga, 2007). Sleep (1976) stated that the boundaries between major world-wide sequences of sediments on continental platforms are due primarily to either eustatic changes or to systematic uplift of the continental interior. If thermal contraction of the lithosphere controls basin subsidence, basins might continue to subside during times of low eustatic sea level. Calculations indicate that significant gaps in the geological record were produced by modest eustatic sea level changes even in rapidly subsiding basins.

A large plume, or "superplume," under the Pacific basin was proposed to account for a number of Cretaceous events, such as a global sea level rise (Hardebeck and Anderson, 1996). According to these authors, the estimated Cretaceous highstand resulting from all the modeled effects is 220–470 m, compared to the observed value of 120–200 m. This discrepancy indicates overestimation of the rate of sea floor and plateau creation and of the size of plume that could have existed. The breakup of the Pangean supercontinent is a more viable explanation of the Cretaceous sea level rise. However, Cloetingh et al. (1985) are of the opinion that no satisfactory tectonic explanations have yet been offered for the apparent sea level fluctuations of about 1 cm/1000 years and a magnitude of up to a few 100 m that have been proposed by Vail et al. (1977).

3.2 EUSTASY AND RELATED CAUSES

Sea level estimates based on sequence stratigraphic data (eg, Haq et al., 1987, 1988) are controversial both because of the proprietary nature of the data on which the estimates were based and because of limitations of the method (eg, Miall, 1991, 2009; Miall and Miall, 2001). Gale et al. (2002) have investigated the sequence stratigraphy of two widely separated marine Cenomanian successions in southeast India and northwest Europe, and used high-resolution ammonite biostratigraphy to demonstrate that the inferred sea level changes are globally synchronous and therefore must be eustatically controlled. Gale et al. (2002) hypothesized that, during pre-Quaternary time, the third-order sequences of Vail and Haq are essentially a sediment response to sea level changes driven by the 400-Ka cycle. The study of Kulpecz et al. (2008) not only illustrated the widely known variability of mixed-influence deltaic systems but also documented the relative stability of deltaic facies systems on the 10^6- to 10^7-year scale, with long periods of cyclically repeating systems tracts controlled by eustasy. Results from the Late Cretaceous showed eustasy as a template for sequences globally; however, regional tectonics (rates of subsidence and accommodation), changes in sediment supply, proximity to sediment input, and flexural subsidence from depocenter loading determine the regional to local preservation and facies expression of sequences.

The sequence stratigraphic data suggest that some very large magnitude third-order sea level changes of 100 m or more occurred (Haq et al., 1987). The only known mechanism for such large, rapid eustatic variations is continental glaciation. Based on a review and comparison with best calibrated Cenozoic-recent data, Boulila et al. (2011) suggested that a range of 25–120 m sea level fluctuation could be possible by glacioeustatic cause. However, accounting continental glaciations as causes may be problematic when they occur during greenhouse climates, such as the Late Cretaceous. The modern pole-to-equator sea level temperature difference is about 50 °C; that of the Mid-Cretaceous ranged from 30 °C to as little as 24 °C, implying a much more equable climate (Hay, 2011), and thus ascribing glacioeustatic cause for high-frequency sea level fluctuations may not be found tenable. However, the review of Boulila et al. (2011) stated that the Mesozoic greenhouse sequences show some relation with the ~2.4-Ma eccentricity cycles, suggesting that orbital forcing contribute to sea level change. These authors also reported that fourth-order sequences may be linked to the astronomically stable 405-Ka eccentricity cycle and to ~160- to 200-Ka obliquity modulation cycles, fifth-order cycles to the short (~100 Ka) eccentricity cycles, and sixth-order cycles to the fundamental obliquity (~40 Ka) and climatic precession (~20 Ka) cycles.

Miller et al. (2003) provided a record of global sea level (eustatic) variations of the Late Cretaceous (99–65 Ma) greenhouse world. Phelps et al. (2015) are of the opinion that the ubiquity of carbonate platforms throughout the Cretaceous Period is recognized as a product of high eustatic sea level and a distinct climatic optimum induced by rapid sea-floor spreading and elevated levels of atmospheric carbon dioxide. Ocean Drilling Program Leg 174AX provided a record of 11–14 Upper Cretaceous sequences in the New Jersey coastal plain that were dated by integrating Sr isotopic stratigraphy and biostratigraphy. Backstripping yielded a Late Cretaceous eustatic estimate for these sequences, taking into account sediment loading, compaction, paleowater depth, and basin subsidence. These authors have also demonstrated that the Late Cretaceous sea level changes were large (>25 m) and rapid (1 Ma), suggesting a glacioeustatic control. Three large $\delta^{18}O$ increases are linked to sequence boundaries, consistent with a glacioeustatic cause and with the development of small ($<10^6$ km^3) ephemeral ice sheets in Antarctica. The sequence boundaries corroborated with sea level falls recorded by Exxon Production Research and sections from northwest Europe and Russia, indicating a global cause, although the Exxon record differed from backstripped estimates in amplitude and shape.

Alsharhan and Kendall (1991) reported contributive effect of tectonics within the eustatically controlled sedimentation in the Late Cretaceous deposits of Abu Dhabi. Gale et al. (2008) attempted a combination of biostratigraphic markers (ammonites, inoceramid bivalves) and carbon isotope excursions to establish a high-resolution correlation between the Middle to Late Cenomanian successions of the Western Interior Basin (USA) and the Anglo-Paris Basin (southern UK). Sequences identified from the sedimentological criteria in the Pueblo succession and elsewhere in the Western Interior Basin coincided precisely with globally recognized sea level events and were thus interpreted to be eustatically controlled. A cross-plot of radiometric ages derived from North American bentonites against an orbitally tuned time scale developed in the Anglo-Paris Basin allowed Gale et al. (2008) to interpret that the sequences were controlled by the 405-Ka long eccentricity cycle. Augmentation of eccentricity cycles by tectonic and feedback mechanism has been reported by Voigt et al. (2006).

Laurin et al. (2005) opined that many ancient rhythmic hemipelagic sequences recorded orbital variations, but the exact nature of the climatic and depositional transfer functions responsible for this link remains poorly understood. Two-dimensional numerical simulations were used by Laurin et al. (2005) to explore selected orbital signal distortion in linked siliciclastic and hemipelagic systems. The models suggested that transfer of multiorder (eg, 20, 100, and 400 Ka) oscillations in relative sea level into the hemipelagic record produced an inherent amplitude distortion of the shorter-period (eg, 20 Ka) cycle.

3.2.1 Eustasy Under Greenhouse Conditions?

During the Cretaceous Period, the climate–ocean system was in flux due to increases in tectonism, global surface and deep-water temperature, atmospheric and oceanic CO_2 concentrations, and Oceanic Anoxic Events. Consequently, the Polar regions may have experienced ice-free summers particularly during the Mid-Cretaceous, resulting in increased sea levels and flooding of interior basins and large variations of sea level (Pugh et al., 2014). Based on the ubiquity of carbonate platforms throughout the Cretaceous Period, Phelps et al. (2015) interpreted them to be the products of high eustatic sea level and elevated levels of atmospheric carbon dioxide. While presenting the revised sea level curve for the Cretaceous, Haq (2014) stated that the causes for these relatively rapid, and at times large amplitude, sea level falls remain unresolved, although based mainly on oxygen-isotopic data, the presence of transient ice cover on Antarctica as the driver remains in vogue as an explanation. Proxies, such as oxygen isotopes, can be used to estimate the volume of water sequestered as glacial ice, but this signal is overprinted by variations in temperature and, to a lesser extent, salinity (Van Sickel et al., 2004). Based on the causes of sea level fluctuations during the Phanerozoic, Miller et al. (2005) stated that sea level mirrors oxygen isotope variations, reflecting ice volume change on the 10^4- to 10^6-year scale, but a link between the oxygen isotope and the sea level on the 10^7-year scale must be due to temperature changes that we attribute to tectonically controlled carbon dioxide variations.

Extended periods of extreme greenhouse conditions (Hunter et al., 2008) and ice-free poles (Hay, 2011) were suggested to have prevalent during the Cretaceous, which negates the possibility of significant sea level fluctuations under the influence of ice volume change, the basic tenet of global eustasy. However, Flögel et al. (2011) suggested large volumes of snow might have accumulated even under Late Cretaceous greenhouse conditions. Integrated chemo-, bio- and sequence stratigraphy from the Cenomanian–Coniacian Monte Turno (Abruzzo, central Italy) allowed Galeotti et al. (2009) a fine-scale calibration of the local sea level curve inferred from the adjacent platform to records of global sea level change. Nevertheless, their results indicated regional marine regressions were coincident with episodes of global cooling and sea level fall, and provided the much needed evidences for a causal link between climate changes and sea level changes. These authors commented that the presence of small polar icecaps during the assumed ice-free Cretaceous is a likely explanation for the observed pattern. Integrated biostratigraphic, chemostratigraphic, sequence stratigraphic, and cyclostratigraphic investigation of Upper Turonian–Lower Coniacian marine strata in the Tethyan Himalaya zone, to retrace the sea level variations and to clarify their global correlations, led Chen et al. (2014) to suggest that Late Turonian–Early Coniacian sea level changes along the southeastern Tethyan margin were controlled by eustasy. The significant regressions during ~90–89.8 Ma and ~92–91.4 Ma, which are recorded in different continents, may be interpreted as the result of continental ice expansion. This observation provided support to the notion of existence of ephemeral polar ice sheets even in the super-greenhouse world. Similarly, through the study of Albian-Cenomanian high-resolution sea level curve of the Western Interior Seaway, USA, and its perfect corroboration with those from other sections located many parts of the World, Koch and Brenner (2009) attributed the observed fluctuations to be the result of eustatic mechanism and concluded that these large-scale and rapid sea level falls during the Mid-Cretaceous greenhouse world to glacioeustasy. They also stated the implication of their study was that a glacioeustatic mechanism operated during the Mid-Cretaceous and also that the definition of a "greenhouse world" as being ice-free needs to be reevaluated. Miller et al. (2004) found synchroneity between the Upper Cretaceous sequence boundaries in New Jersey and the records of Exxon Production Research Company (EPR), northwest European sections, and Russian platform outcrops that pointed to a global cause. Comparison of the eustatic history with published ice-sheet models and Milankovitch predictions suggested that small ($5-10 \times 10^6$ km^3), ephemeral, and a really restricted Antarctic ice sheets paced the Late Cretaceous global sea level change. Over the past 100 Ma, sea level changes reflect global climate evolution from a time of ephemeral Antarctic ice sheets (100–33 Ma).

Perhaps many, if not all, of the "greenhouse" climatic settings included some ice, either in the form of alpine glaciers, small "continental" ice sheets, or both. By varying orbital parameters as well as topography, and atmospheric CO_2 concentrations models, Flögel et al. (2011) indicated the possibility of an Antarctic ice shield build-up large enough to drive sea level fluctuations on the order of tens of meters within ~20,000 years. This is supported under the assumption of pCO_2 levels <800 ppm, low insolation, and elevated topography. The growth of a major Antarctic ice sheet was found possible on reasonable time scales. To accumulate about half the present day snow/ice volume which is required to explain the documented shifts in oxygen isotopes, their model results suggested a time span between 20,000 and 80,000 years for these ice volumes to accumulate. Recently, Lowry et al. (2014) commented that simulations and geological evidence of glaciation and atmospheric composition, showed atmospheric pCO_2 as the primary control on Paleozoic continental-scale glaciation while paleogeographic configurations and solar irradiance were of secondary importance. On the contrary, Eyles (1993) strongly advocated that though significant volumes of ice could have accumulated in Siberia and Antarctica despite significant warm climate, the changes in ice volumes could not have made any marked effect on global sea levels and alternative explanations should perhaps be sought for fourth order, so-called "glacioeustatic" changes in sea level, inferred from the Cretaceous strata.

Miller et al. (2004) concluded that the New Jersey and Russian eustatic estimates are typically one-half of the EPR amplitudes, though this difference varies through time, yielding markedly different eustatic curves. According to Kuhnt et al. (2009), the origins and driving mechanisms of Middle Cretaceous rapid, low-amplitude sea level fluctuations have remained enigmatic. In contrast, low-frequency, high-amplitude sea level changes (second-order sequences) are attributed to tectono-volcanic events such as changes in spreading rates or activity of large igneous provinces. The main body of evidence for Middle Cretaceous rapid sea level fluctuations comes from the identification of regional unconformities (sequence boundaries), resulting from rapid basinward shifts in sediment accumulation due to relative sea level falls. These sea level-related unconformities, which are correlatable over long distances across basins, have been suggested to primarily reflect Middle Cretaceous eustatic sea level fluctuations and their synchroneity with well-studied European basins. However, additional records from different tectonic plates are needed to ascribe Cretaceous sea level fluctuations to eustatic change and to decipher links to potential glaciation events. Miller et al. (2004) were unequivocal based on documenting high-frequency sea level fluctuations in a tectonically stable platform that either continental ice sheets paced sea level changes during the Late Cretaceous, or our understanding of causal mechanisms for global sea level change is fundamentally flawed. Equally unequivocal was the study of Moriya et al. (2007) that presented new higher resolution oxygen isotope records for the Mid-Cenomanian Demerara Rise from extremely well preserved glassy planktonic and benthic foraminifers. These authors opined that the Mid-Cretaceous is widely considered the archetypal ice-free greenhouse interval in Earth history, with a thermal maximum around Cenomanian-Turonian boundary time ($\sim$90 Ma). However, contemporaneous glaciations have been hypothesized based on sequence stratigraphic evidence for rapid sea level oscillation and oxygen isotope excursions in records generated from carbonates of questionable preservation and/or of low resolution. Their records showed no evidence of glaciation, and questioned the hypothesized ice sheets and rendering the origin of inferred rapid sea level oscillations enigmatic. Further, mass-balance calculations made by these authors demonstrated that this Cretaceous sea level paradox is unlikely to be explained by hidden ice sheets existing below the limit of $\delta^{18}O$ detection.

3.2.2 Limno-eustasy

Wagreich et al. (2014) advocated a proposition of "limno-eustasy" to account for sea level fluctuations during ice-free, greenhouse conditions. They argued that the effect of groundwater storage and release on sea level change, particularly important during ice-free greenhouse phases, has been and is widely underestimated in its order of magnitude. It is considered to constitute a water volume that is about equivalent to today's ice volume, thus corresponding to a potential sea level change of up to approximately 50 m applying isostatic adjustment. Groundwater storage, including both freshwater and saline pore waters, strongly exceeds lake and river storage capacities. They introduced the term "limno-eustasy" to describe the effect of water volumes that are bound to groundwater and lake storage on sea level fluctuations and cycles during major greenhouse phases of Earth history. Though seemingly found to be novel and enticing, this proposition has no takers and is met with skeptism owing to many reasons including the absence of any connectivity between inland aquifers to sea, to be capable of transferring water at a rate of reported sea level fluctuations. In addition, no supporting evidences, essentially from outcrop and or deep sea records, as the case that may require, have not been forthcoming either from the proponents or from any other research group anywhere from the world, though hoards of researchers are working on the mechanisms involving eustasy. Thus, limno-eustasy has proved to be a wishful thinking and a nonstarter.

3.2.3 Eustasy Due to Thermal Expansion–Contraction of Sea Water

The increase in volume of matter during heating is known since long. The Earth, being the blue planet, comprising >73% ocean, whose areal distribution (ie, expansion and contraction) in the light of thermal anomalies, especially during the periods of extended periods of global warmth (on a global scale) or major volcanic events, for example, the Deccan Volcanism during Late Cretaceous need to be given due consideration while examining the short-term sea level fluctuations. Though the sea level fluctuation curves and models do consider the expansion and contraction of glacial ice volume, the contribution from thermal expansion–contraction of sea water during the periods of warming–cooling intervals, respectively, and resultant sea level fluctuations in terms of spatial, and temporal scale, rate of change, attenuation effect and relative contribution toward related processes, need to be given adequate consideration, but overlooked so far, perhaps due to the complexity of differentiating this important mechanism from other processes. However, given cognizance to the enormity of the volume of oceanic water, the basic trait of water to expand and contract and its quick response time to respond to thermal events, including climatic warming and volcanic episodes, and resultant changes in the sedimentation and biological systems, this mechanism needs to be studied and methods of recognition of this phenomenon from stratigraphic records need to be developed. The study of Engelhart et al. (2009) is one of the recent works that explicitly recognized the role of thermal expansion of oceanic water as a part of observed sea level rise.

CHAPTER 4

Methods, Tools and Techniques

Reconstruction of past sea level cycles has several implications, including understanding Earth system processes (Engelhart et al., 2011). Over the last few decades, a wide array of methods, tools and techniques has been developed for interpreting sea level change and applied on a variety of terrains and geographic regimes. Yet, there are many pitfalls and difficulties in interpreting sea level change as the sea level change occurs at several time scales simultaneously extending from first-order changes (global tectonics) to fifth-order changes (Milankovitch cyclicity). Another problem is to isolate the tectonic component of sea level change (Munnecke et al., 2010). In an excellent review, Munnecke et al. (2010) detailed the methods of estimating relative sea level fluctuations from stratigraphic records. Citing the work of Johnson (2006), these authors enlisted four types of proxies for interpretation of sea level fluctuation, viz., the sedimentary proxies, the physical proxies, the biological proxies, and the geochemical proxies. A brief description of these tools, their applicability and limitations is presented herein. For more detailed information on specific tools and case studies, readers are advised to consult the papers cited and those listed in them.

4.1 SEDIMENTARY PROXIES

These include traditional facies analysis, outcrop-based, or through geophysical methods in the subsurface (Johnson, 2006). When supplemented with information on bio- or ichnofacies, traditional marine sedimentary facies and facies associations indicate the energy boundaries such as the fair- or storm-weather wave base or the shoreline, and provide a relative sea level curve. Few examples for facies types are depicted in Plates 4.1–4.3. Another example on documentation of stratigraphic variations of facies successions, interpretation of depositional bathymetric variations over time and construction of relative sea

Cretaceous Sea Level Rise. DOI: http://dx.doi.org/10.1016/B978-0-12-805414-7.00004-7

Plate 4.1 (a) Field photograph showing superposition of massive parallel, even bedded moderately sorted typical middle shelf sandstone over deep marine turbiditic deposits. Facies analysis in stratigraphic succession typically indicates the bathymetric variations over time. Deducing the causes of such bathymetric variations is one of the reliable methods of interpreting sea level fluctuations. Location of the photograph: Sillakkudi Formation (Santonian–Campanian)-Karai Formation (Cenomanian) contact in the Cauvery Basin, south India exposed near Kallagam, in the Ariyalur area. (b) Another exposure located adjacent to the previous exposure, showing the fissile, very finely laminated clay deposits of the basinal setting overlain by reverse graded fluvial channel deposits. Interpretation of internal sedimentary structures, in such settings often provides affirmative evidences for sudden (on a geological scale) large scale shoreline shifts and in turn, sea level variation of many tens/hundreds of meters, associated with tectonic forces.

level fluctuation curve is depicted in Fig. 4.1. A step-by-step method of proceeding from documentation of field and facies characteristics, interpretation of depositional processes, environment and setting and finally construction of paleobathymetric/relative sea level fluctuation

Plate 4.2 (a) Cyclic Limestone-Marl alternations, deposited under the influence of Milankovitch scale high-frequency sea level oscillations. Location: Periyakurichchi Quarry section of the Niniyur Formation (Danian), Cauvery Basin, south India. (b) Exposures of beach rocks at the shoreline indicate shoreline retreat towards off-shore, suggestive of sea level decrease. Together with other criteria namely age, tectonic structure, sedimentation dynamics, etc., the causes and rate of such changes could be discerned. Location: River mouth of the Godavari delta, India. (c) Nautilus fossil found embedded on calcareous deposits of Danian age overlying coastal lagoon deposits of Late Maastrichtian. Location: Near Niniyur, Ariyalur area of the Cauvery Basin, south India. (d) Gryphea and alectryonia shells forming a bank in the Maastrichtian limestones. Understanding the lifestyle and environmental preferences of ancient organisms provide an approximation of sea level conditions. Location: Near Kallankurichchi Village, Ariyalur area, Cauvery Basin, South India. (e) Coral colonies of Danian age. Corals are excellent indicators of sea level and its variations, as these are sensitive to bathymetry. Location: Near Niniyur, Ariyalur area, Cauvery Basin.

are depicted in Fig. 4.2, based on data from Cenomanian turbiditic deposits of the Cauvery Basin, South India. Through the documentation of stratigraphic occurrences of oolite, microbialite, and coralgal reef facies types in the Miocene deposits of southeast Spain, Goldstein

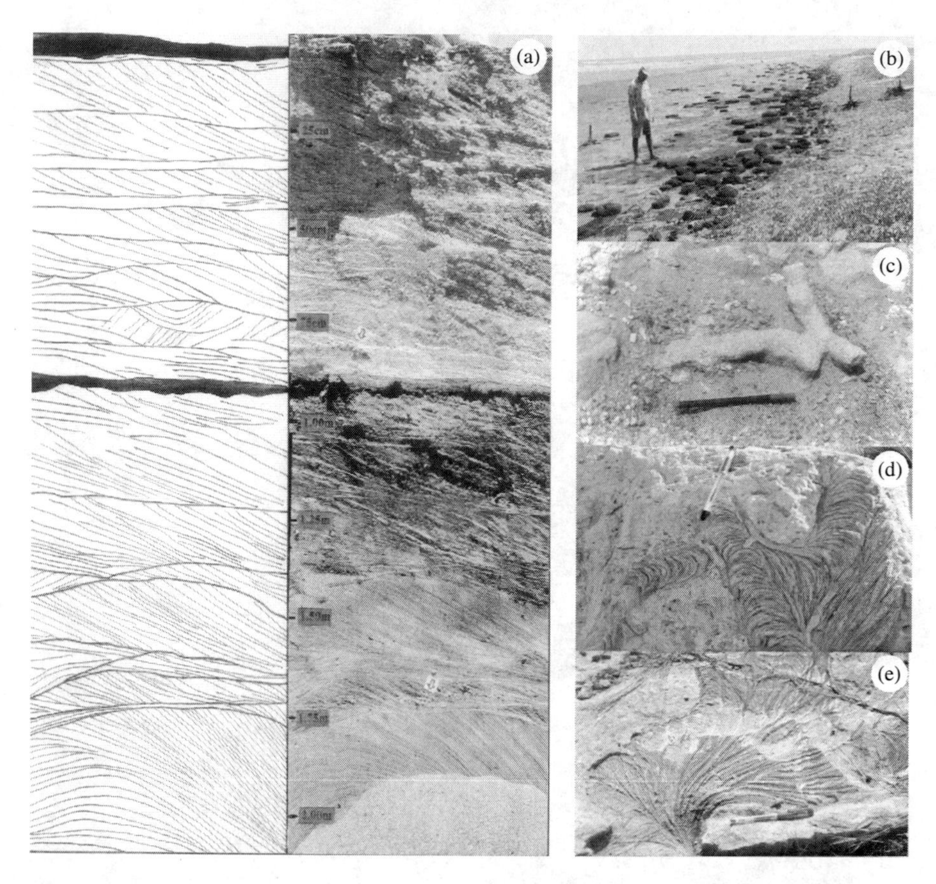

Plate 4.3 Field photographs explaining the utility of integrating facies types and characteristics, internal sedimentary structures, environmental preferences of organisms and ichnofauna for interpreting base level shift, paleobathymetric dynamics, etc. (a) Cliff face of incised estuarine channel, Godavari delta, India, The arrows indicate the occurrences of freshwater fauna, and dark colored massive bed indicates organic matter rich bed. These are exposed in estuarine regime. Change of environment from continental fluvial (as indicated by fauna) to brackish swamp due to sea level fluctuation and ongoing incision due to neotectonic movement are interpreted. (b) The subtidal coastal lagoonal peat deposits, when exposed at supratidal region, offer excellent and affirmative evidences of sea level fluctuation and tectonic land emergence or both, and also provide an estimate of rate of change when combined with absolute age data generated from isotopic methods or biostratigraphic age data. Location of the photograph: Near western mouth of the Krishna delta, India. (c–e) Ichnofauna from the Ariyalur area of the Cauvery Basin, India. These trace fossils offer exclusive insights into the bathymetric variations and are considered as affirmative evidences. See Knaust and Bromley (2012) for a detailed account and Paranjape et al. (2014) for a case study on the effective use of ichnofossils to reconstruct paleobathymetic variations.

et al. (2013) documented the paleotopographic and bathymetric variations, constructed sea level fluctuation pattern, and analyzed the causative mechanism. This study has demonstrated how a simple petrographic and lithofacies mapping on a temporal scale can help interpret complex processes that were operative during geologic past.

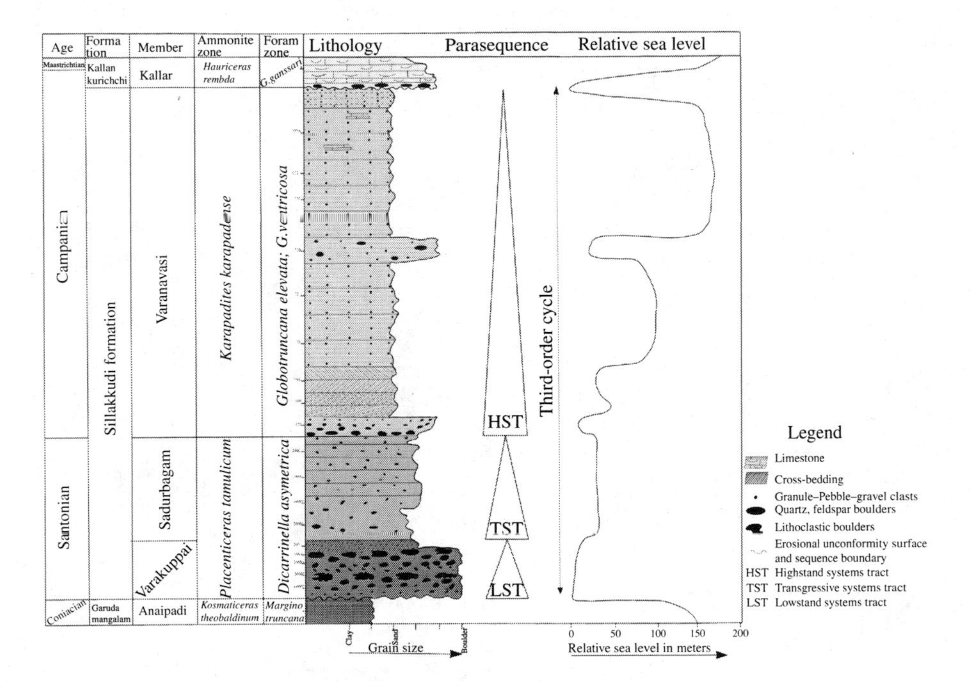

Figure 4.1 ***Facies succession and relative sea level fluctuations of the Santonian–Campanian strata of the Cauvery Basin, South India.*** *From field examination of contact relationships, tectonic structures, ichno and macrofauna, internal sedimentary structures, and laboratory study of texture, microfauna, facies types are established and the depositional bathymetry was interpreted and plotted on a temporal scale.* Source: After Ramkumar et al. (2005a).

There are two ways in which sea level change can be measured; relative to basement ("relative sea level change"—Munnecke et al., 2010) or relative to the center of the Earth ("eustatic sea level change"). A third, commonly used, term is "depositional depth" (local bathymetry), which can change rapidly through autocyclic processes such as delta progradation or even vertical reef platform growth.

Documentation of the facies, sedimentary structures, ichnology, key surfaces, and sandbody and shale geometries found in a marine-to-continental transitional succession at Shivugak Bluffs on the North Slope of Alaska allowed Van Der Kolk et al. (2015) to interpret autogenic and allogenic processes preserved along this ancient Arctic coastline, degree, and magnitude of paleobathymetric trends. High-frequency cycles bundled within Cambrian, Ordovician, Devonian, Jurassic, and Cretaceous greenhouse sequences are characterized by their facies associations and stacking patterns (Elrick and Scott, 2010). Traditionally, these greenhouse stacking patterns have been attributed

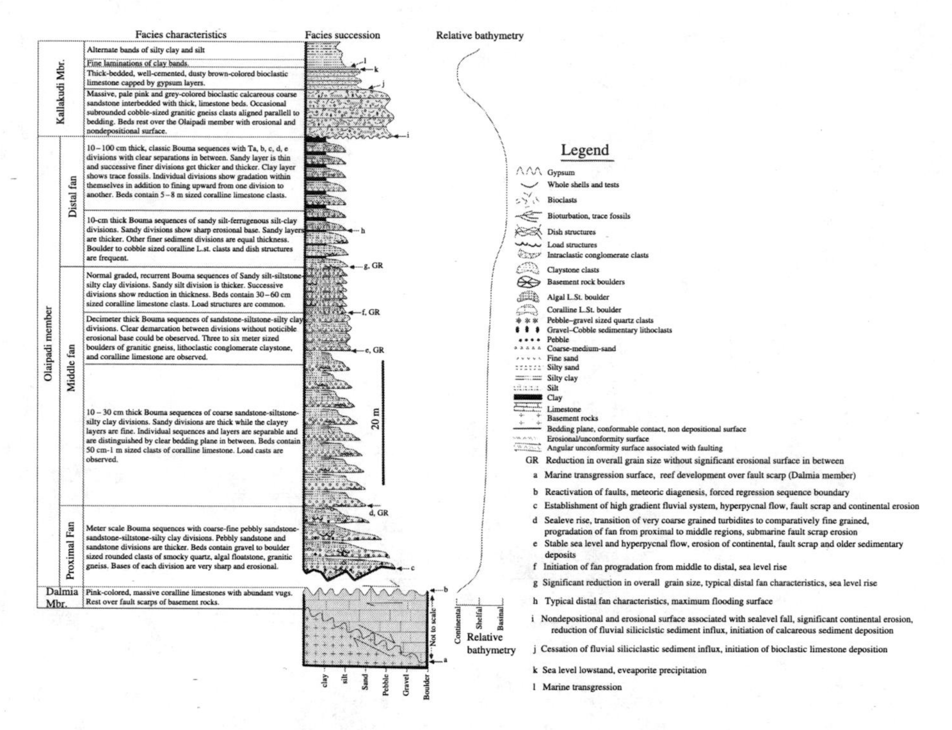

Figure 4.2 ***Facies characteristics, environmental, and depositional bathymetry of the Cenomanian deposits of the Cauvery Basin, South India.*** *Note the distinct facies characteristics at the field scale documented through contact relationships, tectonic and sedimentary characteristics, textural and clast type and morphological traits as explicit in the litholog. Based on these, prevalent environmental processes, depositional setting and their temporal variations are interpreted and translated into paleobathymetric/sea level variations.* Source: After Ramkumar (2008).

to relatively small-magnitude, high-frequency sea level changes superimposed on larger magnitude, Ma-scale sea level changes (Goldhammer et al., 1993). Leckie et al. (2000) demonstrated the use of integrated depositional and paleoecological examination of the lithological units to understand the effect of antecedent paleotopography within the basin at the time of the transgression, documented the variability in sedimentary facies and paleoecology controlled by paleorelief, and demonstrated the diachroneity, lateral extent, and extremity of the multiple unconformities controlled by this paleorelief. After the description of the vertical facies distribution, Bádenas et al. (2003) interpreted the studied succession in terms of long-term relative sea level variation based on the bed-by-bed analysis. Masse et al. (2003) observed that the Late Barremian peritidal platform carbonates of the Marseille region (SE France) show a good correspondence between the inferred relative water depth of facies and their average bed thickness. Cumulative thickness of facies types expresses quantitatively their paleobathymetric range and provides an estimate of their mean water depth. Measuring a number of beds belonging to both shallowing-up and deepening-up sequences, and averaging the results, appear relevant to minimize the effect of compaction, pressure solution and synsedimentary progressive or abrupt changes in accommodation, which are modify bed thickness and its original paleobathymetric significance.

Analysis of the facies pattern and stratigraphic architectures developed during the Cenomanian and Early Turonian based on a bed-by-bed logging of a transect in the Bohemian Massif and the study of onlap patterns allowed quantification of the eustatic sea level rise of the Cenomanian–Turonian Boundary Event (Wilmsen et al., 2010). From an extensive facies analysis of European Callovian–Oxfordian boundary beds, Norris and Hallam (1995) established a bathymetric curve related to a regional sea level cycle. By determining a general correspondence of these events to events on a world-wide scale, a eustatic origin was postulated for the European sea level changes, and a global sea level curve for this time interval was presented. Approximately 1800 m thick Tithonian to Cenomanian carbonates of the Adriatic (Dinaric) platform exposed on the islands of southern Croatia and deposited in a protected and tectonically stable part of the platform interior were studied by Husinec and Jelaska (2006) for the vertical and lateral facies distribution that led to an evaluation of facies dynamics and construction of a relative sea level curve. This curve had showed long-term transgression during the Early Tithonian, Hauterivian, Early Aptian, and Early Albian, resulted in generally thicker beds deposited in subtidal environments of lagoons or shoals. Regression was characterized by shallowing-upward peritidal parasequences, with well-developed tidal-flat laminites commonly capped by emersion breccia and/or residual clay sheets (Early Berriasian, Barremian, Late Aptian, Late Albian). D'Argenio et al. (1997) documented microscale (cm) lithology of Lower Cretaceous (Valanginian–Hauterivian) carbonate platform strata exposed in the Matese Mountains of the southern Apennines, Italy. Based on the vertical organization of lithofacies and their early diagenetic signatures, 46-m scale cycles were recognized, hierarchically categorized into groups (bundles and superbundles) and subjected to spectral analysis. It was found that the studied section was characterized by lithofacies thickness periodicities of 1271, 391, 159, 124, and 69 cm. Linear correlation of the relative ratio set of these periodicities with the corresponding orbital perturbation periodicities showed that 1271 and 391 cm fit the 404,220 and 94,890 year cycles of Early Cretaceous orbital eccentricity; 159 and 124 cm fit the 49,010 and 38,030 year cycles of the Earth's axial obliquity, whereas 69 cm corresponded to the precession periodicity. These authors interpreted that the San Lorenzello section required not <1.6 Ma to form, at an average rate of ~ 26 mm/Ka. Thus, it was concluded that the microstratigraphic approach is very promising for understanding the fine structure of climate-related environmental and sea level fluctuation and for deriving a more precise chronostratigraphy for global correlation. De Vleeschouwer et al. (2013) opined that the interpretation of lithological rhythms as eccentricity cycles should be confirmed by amplitude modulation patterns and by the recognition of precession cycles in stable isotope records.

Brett et al. (1990) attempted field stratigraphic analysis of the Early-Middle Silurian strata of Ontario, New York, and Pennsylvania. They recognized six major sequences, reflecting third-order cycles and observed that each sequence was bounded by widespread unconformities produced by the interplay of eustatic sea level drop and local tectonic uplift. Sequences and component subsequences are correlative basin-wide and are synchronous within the resolution of bio- and event-stratigraphy. Major drops in relative sea level are represented by the sequence-bounding discontinuities. Subsequences are further divisible into widespread minor parasequences and parasequence sets, probably corresponding to sixt and fifth-order sea level cycles, respectively.

Hillgartner and Strasser (2003) attempted to reconstruct a realistic, high-frequency accommodation and sea level curve based on a detailed facies and cyclostratigraphical analysis as employing Fischer plots is not satisfactory, because it does not account for differential compaction, possible erosion, sea level fall below the depositional surface, or subtidal cycles. Their approach included: facies interpretation; cyclostratigraphical analysis and identification of Milankovitch parameters in a well-constrained chronostratigraphic framework; differential decompaction according to facies; estimation of depth ranges of erosion and vadose zone; estimation of water-depth ranges at sequence boundaries and maximum flooding intervals; estimation of mean subsidence rate; classification of depositional sequences according to types of facies evolution: "catch-up," "catch-down," "give-up," or "keep-up"; classification of depositional sequences according to long-term sea level evolution: "rising," "stable," "falling"; calculation of "eustatic" sea level change for each depositional sequence using the parameters inferred from these scenarios; calculation of a sea level curve for each studied section; comparison of these curves among each other to filter out differential subsidence; construction of a "composite eustatic" sea level curve for the entire studied section; and finally, spectral analysis of the calculated sea level curves. Limitations of the method are those common to every stratigraphic analysis.

Harper et al. (2015) are of the opinion that the mixed siliciclastic-carbonate systems do not always behave in a predictable manner, especially in the slope and basin settings. Facies analysis of cores recovered from the Great Barrier Reef margin, Queensland Trough, and Queensland Plateau by these authors revealed that the upper-slope sedimentation resulted from the relationships between sea level fluctuations, antecedent topography, and regional climate that played important role in the type and amount of sediment deposited on the upper slope during glacial, deglacial, and interglacial times. However, though, carbonate–siliciclastic cyclicity as the result of glacioeustatic change with the GBR shelf could be documented, the maximum neritic aragonite export to the upper slope occurred not during peak MIS-5e highstand when sea level was a few meters above modern position, but subsequently during a time (MIS-5d to 5a) when lowered sea level fluctuated between 30 and 50 m below present sea level. Siliciclastic sediments were reworked and exported to the upper slope when the lowstand fluvial plain was re-flooded, whereas neritic carbonate export to the slope reached a maximum when sea level fell and much of the mid to outer shelf reentered the photic zone, subsequent to a drowning interval. These observations made the authors to conclude that mixed margins both modern and ancient would benefit from detailed study of sediment transport in the context of sea level rise and fall.

4.2 PHYSICAL PROXIES

Preserved relief along regional unconformities has provided an independent method namely physical proxy for calibrating and measuring the magnitude of eustatic sea level changes. The method provides direct evidence for the minimum magnitude of sea level change, measured as the distance between the base of the erosional surface and the top of the buried strata (Johnson, 2006). Bond (1978) attempted evaluating the sea level curve through an unconventional tool. Percentages of continental areas that were flooded during a transgression plotted on the corresponding hypsometric curves and used to distinguish between substantial posttransgressive change in continental hypsometries and a transgression caused by a sea level rise followed by little change in continental hypsometries. The method, applied to percentages of flooding during Albian, Late Campanian to Early Maastrichtian, Eocene, and Miocene time intervals, revealed substantial changes in continental hypsometries, especially in Africa, North America, Australia, and Europe.

Van Sickel et al. (2004) employed a method "backstripping," which is a quantitative method of estimating tectonic subsidence, which is defined as the vertical movement of basement in the absence of both sediment loading and sea level change. Miller et al. (2004) estimated sea level variations by backstripping (accounting for paleodepth variations, sediment loading, compaction, and basin subsidence) and found that large (>25 m) and rapid (1 Ma) sea level variations occurred during the Late Cretaceous greenhouse world. Mart-Chivelet (2003) demonstrated a technique, namely, quantitative analysis of accommodation that allowed detailed architectural reconstructions, regional chronostratigraphic correlation, and systems tract interpretation. It was based on several integrated techniques including: construction of total accommodation curves with the aid of backstripping techniques for calculating decompacted sediment accumulation through time, mathematical analysis of these curves and characterization of second- and third-order accommodation patterns, and analysis of parasequence stacking patterns in peritidal cyclic successions by means of Fischer plots. Estimates of eustasy from variations in the changing volume of the oceans, including ridge volumes, sediment volumes, large igneous provinces, continent–continent collision and continental extension are complicated by the fact that subduction removes large portions of the stratigraphic records of sediment volumes and thus, the estimates of older and oldest successions tend to be highly unreliable (Van Sickel et al., 2004).

Goldstein and Franseen (1995) provided a method of constructing sea level curve for the ancient sedimentary records. According to these authors, the quantitative constraints on the history of relative sea level allow for a better understanding of the controls on depositional sequence development. The constraints are provided by "pinning point curves," plots of ancient relative sea level elevations (pinning points) versus time. Constructing a pinning point curve requires identification of ancient stratigraphic positions of sea level through interpretation of facies and surfaces formed at sea level. Then, their ancient relative elevations are determined through reconstructing aspects of ancient paleotopography. Sleep (1976) suggested that the vertical amplitude of eustatic variations can be determined directly from the depth of paleotopographic valleys. The study of Martinson et al. (1998) suggested paleodepth estimation after compaction–correction and incorporating the subsidence data before correlating the sea level changes/curves of geographically separated sections. Their study demonstrated that the recognition of the paleobiotic events and correlatable sea level changes within and beyond the flexural wavelength of foreland basins can be used to distinguish tectonic events from regional sea level changes. The recognition of relatively short-term subsidence events indicated that tectonic events can occur in foreland basins on time scales similar to third-order (or possibly higher) eustatic sea level changes and presented a technique for using high-resolution subsidence analysis to identify asymmetric subsidence. It promises to aid in identifying the tectonic component of subsidence in other foreland basin settings.

Based on the observed vertical dispersion of Holocene sea level data, caused by the superposition of eustatic, isostatic, tectonic, and other factors, Pirazzoli (1993) advocated that similar biases probably exist also for estimations of global sea level change during longer (several Ma) and shorter (1–100 year) geological time periods. He has also suggested that field sea level data are generally only of local value, being affected by a complex of local, regional, and global processes which operate on different time and space scales. Averages or compilations of focal sea level data may lead therefore to misleading estimations of global sea level changes and be biased toward predominant factors that are active in the study areas at the time scale considered. Based on these inferences, he has suggested that, approximate estimations of global sea level changes can be attempted using proxy data, such as variations of oxygen isotope ratios in oceanic core sediments.

4.3 BIOLOGICAL PROXIES

The biological evidence for relative sea level change is based on the depth ranges of fossilized macro or microbiota or their combination in the rocks. Benthic assemblage zones have been widely used for more than four decades, without major conceptual changes, to infer transgressive–regressive cycles (Johnson, 2006). For example, as depicted in Plate 4.2, the nektobenthic organisms (Plate 4.2c) suggest a wider ranges of depth, while the exclusively benthic (Plate 4.2d) suggest more precise values and the sedentary organisms such as coral colonies (Plate 4.2e), divulge a definitive and more acute depth.

Alternations of radiolarian diversity and abundance during Cenomanian–Santonian have been utilized by Pugh et al. (2014) to interpret sea level fluctuations. Krawinkel and Hartmut (1996) utilized sedimentological, paleoecological or taphonomic, and ichnologic criteria to classify stratigraphic bounding surfaces. They demonstrated that the appearance of allochthonous fossil concentrations in certain intervals of a given section, and disappearance in others, is clearly a signal of shallowing or deepening of seas. High-resolution palynoligical analysis has allowed Koch and Brenner (2009) to construct sea level curve for the Mid-Cretaceous deposits of Dakota Formation of Western Interior Seaway, USA, and correlate it with established sea level curves and constrain on the prevalent mechanisms. Martinson et al. (1998) utilized foraminiferal data from seven stratigraphic sections between the Utah-Wyoming thrust belt and western Kansas and recorded 13 correlatable paleoecologic events between Coniacian and Early Campanian time, and provided a framework for interpreting timing of tectonic events and regional sea level changes. On the basis of the review of calcareous nannoplanktons since 225 Ma, Melinte (2004) observed their high environmental fidelity that can be utilized to accurately document paleotemperature, salinity, and sea level variations. Marsh foraminifera demonstrate distinct vertical partitioning where taxon associations can be used with remarkable precision to distinguish between high and low marsh. The relationship between marsh assemblage and elevation with respect to mean sea level is a valuable datum to reconstruct sea level histories. As the average marsh elevation is within approximately 1 m of sea level and that total foraminiferal populations of high and low marsh environments are taxonomically distinct, it is perfectly possible to trace the sea level fluctuations within measurable accuracy. On the basis of this premise, Tibert and Leckie (2004) demonstrated the utility of formaminiferal data in analyzing sea level changes in coastal marine settings. Rossi et al. (2011) demonstrated that foraminiferal assemblages preserved within salt-marsh sediment can provide an accurate and precise means to reconstruct relative sea level due to a strong relationship with elevation, which can be quantified using a transfer function. Transfer functions quantify the relationship between biological assemblages and an environmental variable of interest in the modern environment and can be used to estimate past environmental conditions from paleontological data. High-resolution sea level reconstructions from salt-marsh sedimentary sequences bridge the gap between instrumental records and geological data (Kemp et al., 2009). Accordingly, Rossi et al. (2011) found agreement between the foraminifera-based sea level curve and the tide-gauge records. On the basis of these observations and inferences, they concluded that the transfer function estimates can be extended to sea level reconstructions back into the preinstrumental period as well.

Reuter et al. (2013) observed second-order sea level change superimposed by third-order sea level fluctuations in the Oligocene–Miocene Mediterranean shallow water deposits reflected in changes from planktonic to benthic carbonate production. Notwithstanding the influence of additional important factors on dasycladalean evolutionary history, a strong link between long-term patterns of dasycladalean diversity and global fluctuations in temperature and sea level was recorded by Aguirre and Riding (2005). They also stated that with the exception of the Late Cretaceous, biodiversity broadly tracked temperature from the Carboniferous to the Pliocene. Diversity minima generally corresponded with low sea level and diversity maxima with periods of intermediate sea level. Wendler et al. (2002) carried out a paleoecological reconstruction based on changes in calcareous dinoflagellate cysts (c-dinocysts) assemblages and identified two characteristic ecological assemblages. Gradual changes in absolute and relative abundance of the cyst species in these assemblages over several couplets depicted a bundling pattern that reflected the modulation of the intensity of the precession cycle by the eccentricity cycle (100 Ka). These inferences were further corroborated by independent analyses of the stacking pattern in the natural gamma ray signal and the carbonate and TOC contents. While documenting the cyclic variations of organic carbon and related sea level fluctuations of Cenomanian–Turonian strata of the Tarfaya Atlantic coastal Basin (Morocco), Kuhnt et al. (1997) made an interesting observation of coincidences of high organic burial and benthic fauna free nature. Bachmann et al. (2003) presented a concept combining the stratigraphic study of benthic organisms with graphic correlation and sequence stratigraphy, at a scale that determines the timing of sequence boundaries and quantification of sediment accumulation rates. Both sequence stratigraphic patterns and long-term sea level changes were compared to find similarities to the patterns of Mid-Cretaceous carbonate platform sea level cycles, northern Sinai with those described from other regions.

Climate variability, an essential tenet of eustatic sea level fluctuations, is driven by a complex interplay of global-scale processes and our understanding of them depends on sufficient temporal resolution of the geologic records and their precise interregional correlation, which, in most cases cannot be obtained with biostratigraphic methods alone. The interpretation of past oceanographic events on a supraregional scale requires precisely dated and well-correlated biostratigraphic schemes. Only synchronous events can be interpreted in a global context. Events of local or regional character have therefore to be accurately correlated with time-equivalent shifts in other areas in order to be interpreted in a wider context. One of the problems of interbasin correlation based on biostratigraphy lies in floral and faunal provincialism of the relevant index fossils. In order to overcome such limitations, chemostratigraphy can be used as a stratigraphic tool independent of biostratigraphy (Mutterlose et al., 2014). The combination of chemostratigraphy with biostratigraphy and magnetostratigraphy substantially increases the precision and temporal resolution of interregional correlations and helps overcome problems that arise from differences in biostratigraphic schemes, facies or provincialism of key fossils. Graphic correlation of isotopic shifts, magnetostratigraphic and biostratigraphic events among the sections can detect hiatuses or relative changes of sediment accumulation rates (Wendler, 2013) and can also track the causes of such changes. For example, Voigt et al. (2006) attempted construction of a new high-resolution sea level curve for the Late Cenomanian *Metoicoceras geslinianum*Zone by combining biostratigraphy and carbon isotope stratigraphy of the Anglo-Paris Basin in the United Kingdom and Saxony Basin in Germany.

4.4 GEOCHEMICAL PROXIES

In the last decade, the geochemical proxies have played an increasingly important role in tracking sea level change (Johnson, 2006). Global chemostratigraphic signals such as those carried by organic matter (Middleberg et al., 1991; Pasley et al., 1993; Meyers and Simoneit, 1989; Tu et al., 1999; Calver, 2000), oxygen isotope (Anderson et al., 1996; Veizer et al., 1999), and strontium isotope (Veizer, 1985; Veizer et al., 1999; Mutterlose et al., 2014) and their relationships with sea level changes are well established as a result of research works carried out on a wide variety of terrains and on a spatiotemporal scales (Joo and Sageman, 2014). Occurrences of geochemical and isotopic cycles of varying durations have been unequivocally demonstrated by Veizer et al. (2000) by synthesizing the results covering entire Phanerozoic. Spectral analysis of $\delta^{18}O$ and $\delta^{13}C$ shows that their significant variances are concentrated at 100, 43, 23 and 19 Ka spans (Oppo et al., 1990; Oppo and Fairbanks, 1989). While examining $\delta^{18}O$ of Phanerozoic seawater, Veizer et al. (1997) observed high-frequency cycles within first-order cycle. Strauss (1997) recorded fourth-order cycles of sulfur isotope that stack up to form third-order cycle fluctuations that in turn accommodated within second-order cycles.

Chemostratigraphic correlation based on bulk sediment carbon isotopes is increasingly being used to facilitate high-resolution correlation over large distances, but complications arise from a multitude of possible influences from local differences in biological, diagenetic, and physicochemical factors on individual $\delta^{13}C$ records that can mask the global signal. For employing geochemical proxies to detect sea level fluctuations in the stratigraphic record, separation of the "global signal" from "background noise" is necessary. A global signal is likely to result from varying rates of primary production and carbon burial over time scales longer than the mixing time of the ocean, removing ^{12}C into biomass and ultimately into sediments. Such a signal should be traceable in carbonates that have formed in equilibrium with dissolved CO_2 in the surface layers of seawater and CO_2 in the atmosphere. This appears to have been the case for Cretaceous to recent carbonates, where carbon isotope stratigraphy provides a tool for regional to global correlation but affected by "background noise" due to calcareous plankton formed in ocean surface waters (Brasier et al., 1992).

The global carbon cycle varies on a million year time scale affecting the isotopic and chemical composition of the global carbon (Wallmann, 2001). The glacial intervals coincide with shifts in $\delta^{18}O$ and $\delta^{13}C$. For the carbon isotope record, rate of burial of C_{Org} and thereby changes in atmospheric CO_2 and for the oxygen isotopic records, temperature and ice volume effects on the seawater reservoirs and thereby sea level changes may be linked (Kampschulte et al., 2001). Positive excursions of $\delta^{13}C$ record greater organic productivity and/or organic matter preservation, and decreased carbonate fluxes during periods of rapid sea level rise and the drowning of carbonate platforms. A primary sea level control over the distribution of total organic and carbonate carbon and organic matter type were inferred by White and Arthur (2006) from the Early to Middle Turonian Carlile Formation, Western Interior Basin, United States. However, the high-resolution $\delta^{13}C_{org}$ record along with major- and minor-element proxies, TOC, carbonate content, terrestrial to marine palynomorph ratios, and detailed macro- and microfossil biostratigraphy of the Turonian strata of the Bohemian Cretaceous Basin, Central Europe has not shown any systematic relationship between $\delta^{13}C$ and sea level change (Uličný et al., 2014).

Based on the review of Campanian sea level data from North Africa, the Middle East and northern Europe, Jarvis et al. (2002) observed coincidences of major shifts in $\delta^{13}C$ profiles and eustatic sea level fluctuations. Relatively stable $\delta^{13}C$ values in the Lower Campanian and their long-term fall through the Upper Campanian reflect high and then falling eustatic sea levels, and increased carbonate production. Negative excursions are linked principally to reworking of marine and terrestrial organic matter during rapid sea level fall (Jarvis et al., 2002). Pellenard et al. (2014) performed a very high-resolution carbon and oxygen stable isotope analysis (bulk-carbonate) of a biostratigraphically well-constrained Callovian–Oxfordian series of the eastern Paris Basin. They observed $\delta^{13}C$ and $\delta^{18}O$ fluctuations under the influence of 405 and 100-Ka eccentricity orbital cycles. Uličný et al. (2014) presented a high-resolution $\delta^{13}C_{org}$ record along with major- and minor-element proxies, TOC, carbonate content, terrestrial to marine palynomorph ratios, and detailed macro- and microfossil biostratigraphy. Analysis of the data had shown a number of short-term, basin-wide regressions in the Bohemian Cretaceous Basin, most likely reflecting eustatic falls with recurrence interval of 100 Ka or less.

Kulpecz et al. (2009) interpreted the observed coincidences of increases in $\delta^{18}O$ with sequence boundaries as indications of glacioeustatic sea level fluctuations.

Gröcke et al. (1999) reported that the carbon isotope composition of fossil wood fragments, collected through a biostratigraphically well-constrained Aptian shallow-marine siliciclastic succession on the Isle of Wight, southern Britain, showed distinct variations with time. The results indicated that the stratigraphic signature of $\delta^{13}C_{wood}$ through the Aptian was influenced primarily by fluctuations in the isotopic composition of CO_2 in the global ocean-atmosphere system, as registered in marine carbonates elsewhere, and was not governed by local paleoenvironmental and/or paleoecological factors. Negative and positive excursions in $\delta^{13}C_{wood}$ through the Lower Aptian occur in phase with inferred transgressions and regressions, respectively. As the biofractionation and habitat differences introduce variations in elemental and isotopic compositions, Li et al. (2012) suggested inferring paleotemperature through the use of elemental, elemental ratio and stable isotopic proxies and then interpretation of sea level fluctuations, particularly while selecting sample materials at species level.

The coupling between climate and sea level fluctuations and the sensitivity of geochemical elements to the environmental-climatic and depositional setting allow interpretation and or cross-validation of sea level fluctuation curves. Examination of stratigraphic variations of geochemical compositions of strata under study often helps to interpret sea level fluctuations over time as well as climate of the past. Examples of such studies galore, those have contributed toward this affirmative method of documentation of climatic fluctuations through geochemical proxies. Few of them are briefly presented herein.

Zhao and Zheng (2015) conducted geochemical study of major-trace elements in detrital sediment and carbon–oxygen isotopes in carbonate for a Early Triassic marine stratigraphic profile of Lower Yangtze Basin, China. Their study categorized the geochemical elements into four groups with respect to the difference in their geochemical behaviors. According to these authors, the first group is composed of Al, Th, Sc, Be, In, Ga, K, Rb, and Cs that are tightly correlated due to their immobility during chemical weathering. The second group is composed of Ca and Na that showed anti-sypathetic relationship with Th and Sc, on account of their mobile behavior in the weathering profile. The third group is composed of high field strength elements such as Ti, Nb, Ta, Zr, and Hf that are closely correlated with each other because they were primarily taken up by heavy minerals from sedimentary provenance. The fourth group is composed of redox sensitive elements such as Co, Cu, Fe, Mn, and Ni that are correlated with S and mainly hosted by sulfides. Examination of these behaviors on a temporal scale revealed climatic (warm–cold cycles). A combination of mineral magnetic, geochemical and textural properties of sediments was employed by Basavaiah et al. (2015) to interpret the prevalent climatic-sea level coupling and changes in sea level during Late Quaternary. These authors have conducted conventional carbon dating and integrated the data on stratigraphic variation of magnetic mineral content, textural parameters and palynology to interpret warm–cold climatic intervals and translate them into sea level fluctuations. Similarly, in an innovative approach, Davies et al. (2013) combined the conventional chemostratigraphic signals with that of stratigraphic variation data of mineral magnetic properties, to delineate stratal surfaces that reflect base level shift, ie, shoreline change associated with sea level fluctuation.

Utilization of statistical tools has been in practice since the formative stages of chemostratigraphy as geochemical compositional data are large in volume and need meaningful reduction without losing the basic traits of the strata under study. A variety of parametric and nonparametric tools namely, correlation, correspondence, ANOVA, cluster, factor, and discriminant function analyses are employed for this purpose. Montero-Serrano et al. (2010) introduced a statistical protocol for recognition and characterization of chemofacies. It involved data preparation and cleaning: selection of relevant components, convenient replacement of values below the detection limit and determination of outliers. It is followed by a biplot analysis to infer geochemical processes that can be interpreted from a paleoenvironmental point of view: detrital association, redox–organic matter association and carbonate association. Considering such geochemical associations, a constrained cluster analysis has to be then carried out to determine the chemofacies for each section. According to the compositional nature of geochemical data, all the statistical analyses have to be conducted within a log-ratio analysis framework. In addition, robust statistical methods have to be considered for outlier detection and biplot representation in order to smooth the influence of potential outliers on the estimates. Hilgen et al. (2014) suggested employing prudence instead of strict compliance and adherence of statistical confidence levels in analyzing the time series of long-range cyclic data for effective modeling. Cahill et al. (2015) have used the foraminiferal and stable isotopic data to construct sea level fluctuations through a complex statistical protocol and demonstrated that reconstructing the continuous and dynamic evolution of relative sea level change with fully quantified uncertainty is possible.

4.5 INTEGRATED ANALYSES FOR DOCUMENTING SEA LEVEL FLUCTUATIONS

Reuter et al. (2013) stated that though the shallow water deposits are sensitive to environmental settings, they may not be readily suitable for global correlation as do the deeper water deposits. In order to obviate this lacuna, these authors suggested use of multiproxy namely, facies analyses, $CaCO_3$ and total organic carbon (TOC) contents and stable carbon isotope records. Galeotti et al. (2009) conducted integrated chemo-, bio- and sequence stratigraphic study on the Cenomanian–Coniacian succession of Monte Turno (Abruzzo, central Italy) to develop a fine scale calibration of the local sea level curve inferred from the adjacent platform to records of global sea level change. Li et al. (2000), Adatte et al. (2002), and Stüben et al. (2002) employed multi-disciplinary study of sea level and climate proxies, including bulk rock and clay mineral compositions, carbon isotopes, TOC, Sr/Ca ratios, and macro- and microfaunal associations. Their approach revealed seven major sea-level regressions in the southwestern Tethys during the last 10 million years of the Cretaceous. These sea-level changes were interpreted as eustatic. These authors reasoned that the climatic changes inferred from clay mineral contents correlate with sea level changes: warm or humid climates accompany high sea levels and cooler or arid climates generally accompany low sea levels.

Investigations of 13 Turonian–Maastrichtian sections of central-east Sinai, for facies zones, biostratigraphy (planktonic foraminifera, calcareous nannofossil, ostracoda and ammonite) and paleoecology and a review of all published data led Luening et al. (1998) to construct a sea level curve and correlate it with coeval sections located in Egypt, Israel, Jordan and Tunisia and find synchroneity and eustatic origin for them. Jowett et al. (2007) attempted outcrop sedimentary analysis along with foraminiferal micropaleontology and Rock-Eval pyrolysis to investigate the relative sea level. Based on the changes in the mineralogical composition namely, quartz content and clay mineral assemblage, and associated variation of the major-element chemistry, Hofmann et al. (2001) were able to infer cyclic climatic changes combined with flooding of coastal lowlands during an overall transgressive phase in the Early Albian in northern Africa. Miller et al. (2004) developed a Late Cretaceous sea level estimate from Upper Cretaceous sequences at Bass River and Ancora, New Jersey (Ocean Drilling Program Leg 174AX). These authors have dated the recognized sequences by integrating Sr isotope and biostratigraphy (age resolution ± 0.5 Ma) and then estimated paleoenvironmental changes within the sequences from lithofacies and biofacies analyses. Kulpecz et al. (2008) integrated sequences defined on the basis of lithologic, biostratigraphic, and Sr-isotopic stratigraphy from cores with geophysical log data from 28 wells to further develop and extend methods and calibrations of well-log recognition of sequences and facies variations. These authors were able to recognize and correlate the sequences at high resolution (>1 Ma). Bádenas et al. (2012) suggested the use of sedimentological, lithologic, and chemostratigraphic (Mg/Ca, $\delta^{13}C$, $\delta^{18}O$) analysis for characterization of high-frequency cycles (centimeter- to decimeter-thick depositional units). Ghirardi et al. (2014) employed spectral analyses, using the multitaper and the amplitude spectrogram methods to detect sedimentary cycles related to an orbital forcing.

In an innovative study, Van der Meer et al. (2014) recently demonstrated the control exercised by tectonics over atmospheric CO_2 through a complex and intrinsically coupled chain of processes and thereby over climate–continental weathering and sea level fluctuations, and ensuing sedimentary records. Hiscott (2001), Spalletti et al. (2001), Bachmann et al. (2003), and Brett et al. (2004) opined that it is a common phenomenon of sedimentary records to have sea level cycles affected by local tectonics either positively or negatively. Depending on the local, regional and global scale of processes, these cyclic changes may get preserved in the ensuing sedimentary strata, the temporal scale of which may vary from few thousand years to few or few tens of millions of years – a postulate widely utilized in seqeuence and chemostratigraphy. Thus, the enigma of relative roles of tectonics-sea level fluctuations over depositional pattern remains to be there where it had all started when Vail et al. (1977) proposed the sequence-stratigraphic concepts. It has also raised questions on the very fundamentals of sequence and chemostratigraphic applications. At this juncture, it becomes essential to address the problem of discrimination relative influences of tectonics and relative sea level fluctuations over sedimentary records. The recent study of Zorina (2014) is one of the good examples that demonstrated the importance of discriminating eustatic and tectonic signals.

Ramkumar (2015) had shown the occurrences of two second-order cycles containing six third-order cycles, within which a multitude of fourth and higher order cycles of sea level fluctuations in the Cauvery Basin, India. It is interesting to note that the paleobathymetric fluctuation chart (Fig. 4.3) of Barremian–Danian constructed by Ramkumar et al. (2004) based on lithofacies succession had detected and documented the occurrences of all the six global eustatic peaks of the duration, recorded independently through foraminiferal data (Raju and Ravindran, 1990; Raju et al., 1993), which in turn were affirmed through statistically distinct parasequences and chemozones (Ramkumar et al., 2011). The chemozones, definable in terms of detrital influx dominated Si-group of

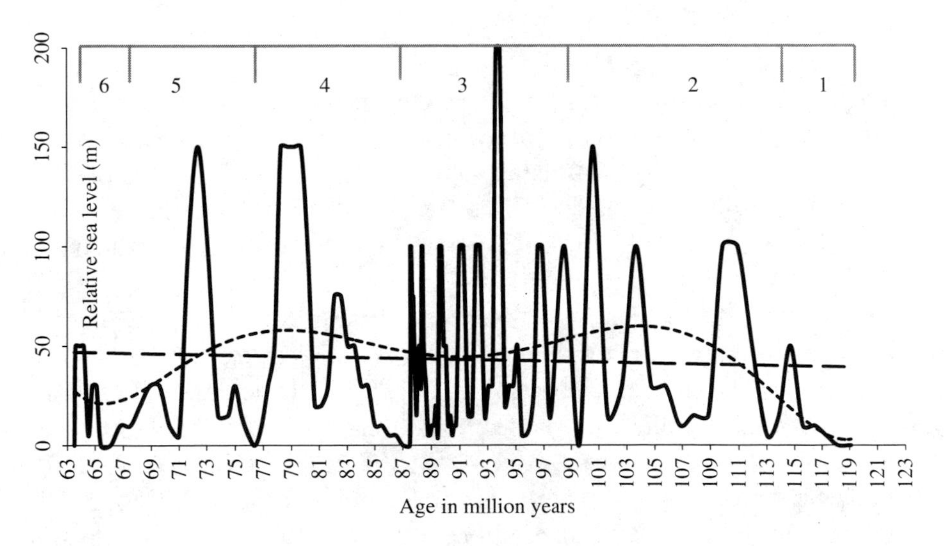

Figure 4.3 ***Showing bathymetric variations during Barremian–Danian in the Cauvery Basin, India.*** *The solid black line (solid line in print versions) represents the absolute bathymetric variations, the dashed green (fine-dashed in print versions) line depicts the linear trend, and the dashed blue (long-dashed in print versions) line indicates the polynomial trend. While the peaks in the polynomial trend indicate two second-order cycles within which six third-order cycles, multiple fourth and higher order cycles as represented by absolute value curve can be recognized.* Source: After Ramkumar et al. (2004).

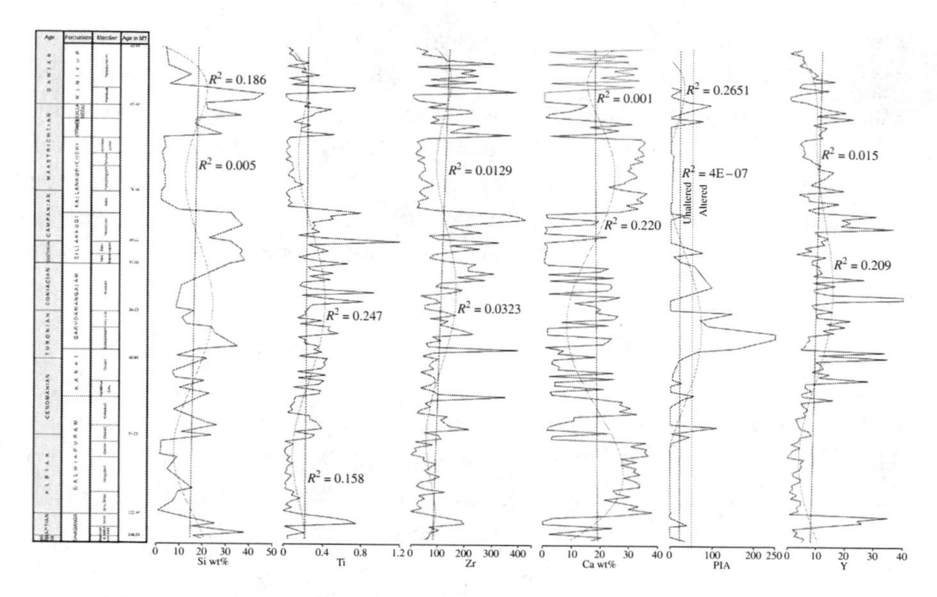

Figure 4.4 ***Temporal variations of geochemical and other proxies of Barremian–Danian strata of the Cauvery Basin, India.*** *The solid line indicates absolute values, the long-dashed line indicates the linear trend, and the fine-dashed line depicts the polynomial trend. Compare the hierarchy of the occurrences of second, third, and higher order fluctuations in tune with the bathymetric trends depicted in Fig. 4.1. Note that, the trends in the profiles of Si-Ti-Zr and Ca are always antisympathetic and are also reflected in Plagioclase index of Alteration (PIA) profile, signifying climate-sea level controlled nature. The short-sharp positive anomalies of* Y *are associated with tectonic events in the basin. Employing multiproxies is thus, leads to accurate and cross-validated bathymetric trends.* Source: Compiled from Ramkumar (2015) and Ramkumar et al. (2011).

elemental concentrations and biogeochemical sedimentation dominated Ca-group of elemental concentrations have shown a synchrony of climatic variations (Fig. 4.4) as explicit in temporal variation of plagioclase index of alteration (PIA). These results also emphasized the utility of employing independently verifiable, multiproxy studies for deducing a hierarchy of sea level fluctuations. Integration of field scale facies characteristics, tectonic and sedimentary structures, micro and macrofaunal assemblages, textural properties of the rocks, environmental preferences of fossils, mineralogical, major, trace elemental, and carbon, oxygen and strontium isotopic compositional trends were utilized to understand the geological events of the Late Cretaceous–Early Paleogene strata of the Cauvery Basin (Ramkumar et al., 2005b, 2010) and compared the sea level fluctuations at local, regional and global scale (Fig. 4.5), that forms one of the examples of utilization of multiproxy studies. These studies have also established global scale processes that occurred independently of each other, but synchronously.

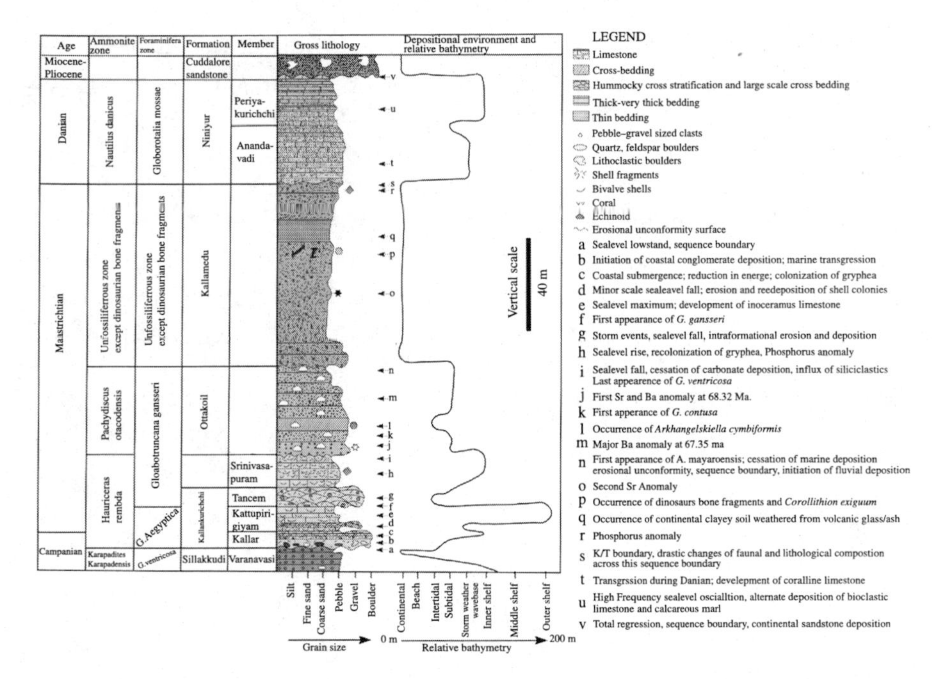

Figure 4.5 ***Multiproxy characterization of K/Pg section of the Cauvery Basin, South India.*** *Field scale tectonic and sedimentary structures, facies characteristics, contact relationships, occurrence and abundance of mega, micro and nanno biota, petrographic, mineralogic, and geochemical data were utilized to characterize and cross validate the prevalent sea level fluctuations.* Source: Modified after Ramkumar et al. (2010).

Sensitivity of geochemical proxies, their susceptibility to local influences and invisibility to get amalgamated into global signals have been documented by the recent study of Qie et al. (2015). These authors employed integrated study of the litho-, bio-, and chemostratigraphy of the Devonian–Carboniferous boundary at four sections (Qilinzhai, Malanbian, Gedongguan and Long'an) in South China to understand paleoenvironmental changes as enforced by sea level fluctuations and controls on $\delta^{13}C_{carb}$ variation. All the four studied profiles have shown positive $\delta^{13}C_{carb}$ shift, suggesting a biological pump that existed in the aftermath of the latest Devonian glaciation. However, peak $\delta^{13}C_{carb}$ values differ markedly among the studied sections, suggesting that local carbon cycling processes played an important role. Engelhart et al. (2011) developed a detailed methodology to reconstruct relative sea level.

The microfauna, especially foraminifera, like many other benthic inhabitants of marine habitats, are highly sensitive to several physical, chemical and several other oceanographic parameters such as nature and stability of substrate, bathymetry, energy, salinity, oxygen content, dissolved organic carbon content, etc. These traits are successfully exploited by integrated analysis of sediment properties and faunal content, to accurately estimate the paleobathymetric oscillations. There are few large databases established for this purpose. Recently, Rossi and Horten (2009) have evaluated the efficiency and accuracy of one such database to find that it is possible to reconstruct paleobathymetric variations based on integrated sedimentological and foraminiferal data, and the predictability of the database is more accurate ($R^2 = 0.95$) in depth range between 8 and 100 m.

CHAPTER 5

Where It Stands and Where Is It Headed

Significant advancements have been made during the past few decades in terms of improvements in very-high resolution chrono-, bio-, magneto-, sequence, and chemostratigraphic datum and the same are calibrated with astronomic time scales. These have significantly enhanced the possibility of correlating and dating the short-term Cretaceous sea level records with dependable accuracy. In addition, with these improvements it is becoming more realistic than what was possible few decades ago. Wider acceptance and applicability has been achieved especially, with reference to the resolution, and scale of correlation, than what was achievable when the sequence concepts were introduced and invigorated the sea level research or, the reports of OAE'S and CORB'S in the Cretaceous strata were published and followed by a host of similar publications through international collaborative research initiatives or, the never-ending discussions commenced on the end-Cretaceous events that led to significant dwindling and extinction of biota. In this context, the IGCP Project 609 intends to differentiate and quantify the short- and long-term sea level records of Mid- to Late Cretaceous (120–66 Ma) within the newly available high-resolution absolute time scale based on orbital cyclicity. The time interval for the intended study begins with the first major oceanic anoxic event (OAE 1a) and terminates at the end of the Cretaceous. It includes the time of super-greenhouse conditions, the major oceanic anoxic events, the Cretaceous thermal maximum and the subsequent cooling to ordinary greenhouse conditions and also the time period of end-Cretaceous during which catastrophic and noncatastrophic events took place leading to one of the important extinction events of the Earth's history.

The review presented under different chapters and sections in this book makes it known that our current understanding on the Cretaceous sea level cycles is far from complete. Though sea level highs and lows are recorded in widely separated basins and are linked

Cretaceous Sea Level Rise. DOI: http://dx.doi.org/10.1016/B978-0-12-805414-7.00005-9

to the eustasy, consensus on the causative mechanisms is yet to be evolved. It was principally due to the fact that despite sea level fluctuations during the Cretaceous are cyclic, they are only periodic. Added to this are the problems of poorly constrained strata, gaps in sedimentary records, lack of appreciation in utilizing multiproxies, complexities introduced by local and regional tectonic and other processes within the eustatic cycles and above all, lack of consensus on the existence and role of glacial feedback mechanism during the greenhouse conditions. These complexities and lack of consensus on the mechanisms need to be addressed especially through the application of multiproxy studies on selected extended stratigraphic records.

There is yet another issue that needs to be addressed. The perturbations among the cyclic nature of Cretaceous sea level cycles have been the focus of many studies, yet consensus on the causes and consequences elude the researchers. Notable among the perturbations are those linked with oceanic anoxia during early part of Upper Cretaceous and those linked with the dwindling of species diversity/population, environmental stress and extinction during the Late Cretaceous. Though the scientific community could document suitable stratigraphic sections representing marine and nonmarine realms of these time slices, concerted efforts to characterize the strata through multiproxy studies are awaited.

The following summarize the past, present, and future of Cretaceous sea level research.

- The Cretaceous, particularly the time span from 120 to 65 Ma is one of the very important time slices of the Earth's history. This duration had marked extensive volcanism, drastic changes in climatic conditions, ocean-basin volume, biodiversity, etc., and hence has been the subject matter of studies by the scientific communities.
- Nevertheless, owing to the many limitations including lack of appreciation for the application of multiproxy studies and the embedded and coeval nature of different causes that contributed toward fewer consequences, consensus eludes the scientific community. More so is on the short-term sea level cycles, their causative mechanisms and the consequences. The UNESCO IGCP–609 intends alleviating these impediments through facilitation of international collaborative research with clear mandates and it looks bright at the end of the tunnel.
- As it lies at the intersection of Earth's solid, liquid, and gaseous components, the sea level links the dynamics of the fluid part of the planet

with those of the solid part of the planet and hence, multiple processes exert varying degrees of influences on the sea level, which varies simultaneously on temporal scales of first order to Milankovitch band and in turn shows imprints of local, regional, and global processes. Hence, developing reliable tools and methods for accurately discriminating the imprints of different processes, degrees of influences, and the temporal scales is necessary.

- Evolving consensus on the applicability of different tools, followed by designation of selected stratigraphic sections for specified time slices, short-term cycles, events, etc., and conductance of collaborative multiproxy studies hold the key for making significant advancement of our understanding on the causes and consequences of sea level changes.

REFERENCES

Adatte, T., Keller, G., Stinnesbeck, W., 2002. Late Cretaceous to early Paleocene climate and sea level fluctuations: the Tunisian record. Palaeogeogr. Palaeoclimatol. Palaeoecol. 178, 165–196.

Aguirre, J., Riding, R., 2005. Dasycladalean algal biodiversity compared with global variations in temperature and sea level over the past 350 Myr. Palaios 20, 581–588.

Alsharhan, A.S., Kendall, C.G.St.C., 1991. Cretaceous chronostratigraphy, unconformities and eustatic sea level changes in the sediments of Abu Dhabi, United Arab Emirates. Cretaceous Res. 12, 379–401.

Anderson, J.B., Abdulah, K., Sarzalejo, S., 1996. Late quaternary sedimentation and high-resolution sequence stratigraphy of the east Texas shelf. In: De Batist, M., Jacobs, P. (Eds.), Geology of Siliciclastic Shelf Seas, vol. 117. Geological Society Special Publication., pp. 95–124.

Armitage, J.J., Duller, R.A., Whittaker, A.C., Allen, P.A., 2011. Transformation of tectonic and climatic signals from source to sedimentary archive. Nat. Geosci. Available from: http://dx.doi.org/10.1038/ngeo1087.

Bachmann, M., Hirsch, F., 2006. Lower Cretaceous carbonate platform of the eastern Levant (Galilee and the Golan Heights): stratigraphy and second-order sea level change. Cretaceous Res. 27, 487–512.

Bachmann, M., Amin, M., Bassiouni, A., Kuss, J., 2003. Timing of Mid-Cretaceous carbonate platform depositional cycles, northern Sinai, Egypt. Palaeogeogr. Palaeoclimatol. Palaeoecol. 200, 131–162.

Bádenas, B., Aurell, M., Rodríguez-Tovar, F.J., Pardo-Igúzquiza, E., 2003. Sequence stratigraphy and bedding rhythms of an outer ramp limestone succession (Late Kimmeridgian, Northeast Spain). Sediment. Geol. 161, 153–174.

Bádenas, B., Aurell, M., Armendáriz, M., Rosales, I., García-Ramos, J.C., Piñuela, L., 2012. Sedimentary and chemostratigraphic record of climatic cycles in Lower Pliensbachian marl–limestone platform successions of Asturias (North Spain). Sediment. Geol. 281, 119–138.

Basavaiah, N., Babu, J.L.V.M., Gawali, P.B., Kumar, K.Ch.V.N., Demudu, G., Prizomwala, S.P., et al., 2015. Late quaternary environmental and sea level changes from Kolleru Lake, SE India: inferences from mineral magnetic, geochemical and textural analyses. Quat. Int. 371, 197–208.

Bond, G., 1978. Speculations on real sea level changes and vertical motions of continents at selected times in the Cretaceous and Tertiary Periods. Geology 6, 247–250.

Boulila, S., Galbrun, B., Miller, K.G., Pekar, S.F., Browning, J.V., Laskar, J., et al., 2011. On the origin of Cenozoic and Mesozoic "third-order" eustatic sequences. Earth Sci. Rev. 109, 94–112.

Brasier, M.D., Anderson, M.M., Corfield, R.M., 1992. Oxygen and carbon isotope stratigraphy of Early Cambrian carbonates in southeastern Newfoundland and England. Geol. Mag. 129, 265–279.

Brett, C.E., Goodman, W.M., LoDuca, S.T., 1990. Sequences, cycles, and basin dynamics in the Silurian of the Appalachian Foreland Basin. Sediment. Geol. 69, 191–244.

Brett, C.E., McLaughlin, P.I., Cornell, S.R., Baird, G.C., 2004. Comparative sequence stratigraphy of two classic Upper Ordovician successions, Trenton Shelf (New York–Ontario) and Lexington Platform (Kentucky–Ohio): implications for eustasy and local tectonism in eastern Laurentia. Palaeogeogr. Palaeoclimatol. Palaeoecol. 210, 295–329.

Cahill, N., Kemp, A.C., Horton, B.P., Parnell, A.C., 2015. Modeling sea-level change using errors in-variables integrated gaussian processes. Ann. Appl. Stat. 9, 547–571.

Calver, C.R., 2000. Isotope stratigraphy of the Ediacarian (Neoproterozoic III) of the Adelaide rift complex, Australia and the overprint of water column stratification. Precambrian Res. 100, 121–150.

Carter, R.M., Abbott, S.T., Fulthorpe, C.S., 1991. Application of global sea level and sequence stratigraphic models on Southern hemisphere Neogene strata from New Zealand. Int. Assoc. Sedimentol. Spec. Publ. 12, 41–65.

Catuneanu, O., 2006. Principles of Sequence Stratigraphy. Elsevier B.V., Amsterdam.

Catuneanu, O., Galloway, W.E., Kendall, C.G.St.C., Miall, A.D., Posamentier, H.W., Strasser, A., 2011. Sequence stratigraphy: methodology and nomenclature. Newsl. Stratigr. 44, 173–245.

Chen, X., Wang, C., Wu, H., Kuhnt, W., Jia, J., Holbourn, A., et al., 2014. Orbitally forced sea level changes in the Upper Turonian-Lower Coniacian of the Tethyan Himalaya, southern Tibet. Cretaceous Res., 1–11. Available from: http://dx.doi.org/10.1016/j.cretres.2014.07.010.

Cloetingh, S., Haq, B.U., 2015. Inherited landscapes and sea level change. Science 347, 393–404.

Cloetingh, S., McQueen, H., Lambeck, K., 1985. On a tectonic mechanism for regional sea level variations. Earth Planet. Sci. Lett. 75, 157–166.

Cogné, J., Humler, E., Courtillot, V., 2006. Mean age of oceanic lithosphere drives eustatic sea level change since Pangea breakup. Earth Planet. Sci. Lett. 245, 115–122.

Conrad, C.P., 2013. The solid Earth's influence on sea level. Geol. Soc. Am. Bull. 125, 1027–1052.

D'Argenio, B., Ferreri, V., Amodio, S., Pelosi, N., 1997. Hierarchy of high-frequency orbital cycles in Cretaceous carbonate platform strata. Sediment. Geol. 113, 169–193.

Davies, E.J., Ratcliffe, K.T., Montgomery, P., Pomar, L., Ellwood, B.B., Wray, D.S., 2013. Magnetic Susceptibility (χ) Stratigraphy and Chemostratigraphy Applied to an Isolated Carbonate Platform Reef Complex, Llucmajor Platform, Mallorca, vol. 105. Society of Sedimentary Geology Special Publication. Available from: http://dx.doi.org/10.2110/sepmsp.105.05

De Vleeschouwer, D., Rakocinski, M., Racki, D., Bond, D.P.G., Sobien, K., Claeys, P., 2013. The astronomical rhythm of Late-Devonian climate change (Kowala section, Holy Cross Mountains, Poland). Earth Planet. Sci. Lett. 365, 25–37.

Elrick, M., Scott, L.A., 2010. Carbon and oxygen isotope evidence for high-frequency (104–105 yr) and My-scale glacio-eustasy in Middle Pennsylvanian cyclic carbonates (Gray Mesa Formation), central New Mexico. Palaeogeogr. Palaeoclimatol. Palaeoecol. 285, 307–320.

Engelhart, S.E., Horton, B.P., Douglas, B.C., Peltier, W.R., Törnqvist, T.E., 2009. Spatial variability of Late Holocene and 20th century sea-level rise along the Atlantic coast of the United States. Geology 37, 1115–1118.

Engelhart, S.E., Horton, B.P., Kemp, A.C., 2011. Holocene sea level changes along the United States' Atlantic Coast. Oceanography 24, 70–79.

Esmeray-Senlet, S., Özkan-Altiner, S., Altiner, D., Miller, K.G., 2015. Planktonic foraminiferal biostratigraphy, microfacies analysis, sequence stratigraphy, and sea-level changes across the Cretaceous–Paleogene boundary in the Haymana Basin, Central Anatolia, Turkey. J. Sediment. Res. 85, 489–508.

Eyles, N., 1993. Earth's glacial record and its tectonic setting. Earth Sci. Rev. 35, 1–248.

Flögel, S., Wallmann, K., Kuhnt, W., 2011. Cool episodes in the Cretaceous—exploring the effects of physical forcings on Antarctic snow accumulation. Earth Planet. Sci. Lett. 307, 279–288.

Gale, A.S., Hardenbol, J., Hathway, B., Kennedy, W.J., Young, J.R., Phansalkar, V., 2002. Global correlation of Cenomanian (Upper Cretaceous) sequences: evidence for Milankovitch control on sea level. Geology 30, 291–294.

Gale, A.S., Voigt, S., Sageman, B.B., Kennedy, W.J., 2008. Eustatic sea level record for the Cenomanian (Late Cretaceous)—extension to the Western Interior Basin, USA. Geology 36, 859–862.

Galeotti, S., Rusciadelli, G., Sprovieri, M., Lanci, L., Gaudio, A., Pekar, S., 2009. Sea level control on facies architecture in the Cenomanian–Coniacian Apulian margin (Western Tethys): a record of glacio-eustatic fluctuations during the Cretaceous greenhouse?. Palaeogeogr. Palaeoclimatol. Palaeoecol. 276, 196–205.

Ghirardi, J., Deconinck, J., Pellenard, P., Martinez, M., Bruneau, L., Amiotte-Suchet, P., et al., 2014. Multi-proxy orbital chronology in the aftermath of the Aptian Oceanic Anoxic Event 1a: palaeoceanographic section, Vocontian Basin, SE France. Newsl. Stratigr. 47, 247–262.

Gjelberg, J., Steel, R.J., 1995. Helvetiafjellet Formation (Baremian-Aptian) Spitsberen: Characteristics of a Transgressive Succession, vol. 5. NPF Special Publication.

Goldhammer, R.K., Oswald, E.J., Dunn, P.A., 1991. Hierarchy of stratigraphic forcing: example from Middle Pennsylvanian shelf carbonates of the Paradox basin. In: Franseen, E.K., Watney, W.L., Kendall, C.G.St.C. (Eds.), Sedimentary Modeling, vol. 233. Kansas Geological Survey Bulletin, pp. 361–413.

Goldhammer, R.K., Lehmann, P.J., Dunn, P.A., 1993. The origin of high-frequency platform carbonate cycles and third-order sequences (Lower Ordovician El Paso Gp, West Texas): constraints from outcrop data and stratigraphic modeling. J. Sediment. Petrol. 63, 318–359.

Goldstein, R.H., Franseen, E.K., 1995. Pinning points: a method providing quantitative constraints on relative sea level history. Sediment. Geol. 95, 1–10.

Goldstein, R.H., Franseen, E.K., Lipinski, C.J., 2013. Topographic and sea level controls on oolite-microbialite-coralgal reef sequences: the terminal carbonate complex of Southeast Spain. Am.Assoc. Petroleum Geol.Bull. 97, 1997–2034.

Grammer, G.M., Eberli, G.P., Van Buchem, F.S.P., 1996. Application of high resolution sequence stratigraphy to evaluate lateral variability in outcrop and subsurface—Desert Creek and Ismay intervals, Paradox basin. In: Longman, M.W., Sonnelfeld, M.D. (Eds.), Paleozoic Systems of the Rocky Mountain Region, Rocky Mountain Section. Society of Economic Palaeontologists and Mineralogists, pp. 235–266.

Gröcke, D.R., Hesselbo, S.P., Jenkyns, H.C., 1999. Carbon-isotope composition of Lower Cretaceous fossil wood: ocean-atmosphere chemistry and relation to sea level change. Geology 27, 155–158.

Groetsch, J., 1996. Cycle stacking and long-term sea level history in the Lower Cretaceous (Gavrovo Platform, NW Greece). J. Sediment. Res. 66, 723–736.

Hallam, A., Wignall, P.B., 1999. Mass extinctions and sea level changes. Earth Sci. Rev. 48, 217–250.

Hancock, J.M., 1993. Comments on the EXXON cycle chart for the Cretaceous system. Cuadernos de Geologia Iberia 17, 57–78.

Hancock, J.M., Kauffman, E.G., 1979. The great transgressions of the Late Cretaceous. J. Geol. Soc. 136, 175–186.

Haq, B.U., 2014. Cretaceous eustasy revisited. Glob. Planet. Change 113, 44–58.

Haq, B.U., Hardenbol, J., Vail, P.R., 1987. Chronology of fluctuating sea levels since the Triassic. Science 235, 1156–1167.

Haq, B.U., Hardenbol, J., Vail, P.R., 1988. Mesozoic and Cenozoic chronostratigraphy and cycles of sea level change. In: Wilgus, C.K., Hastings, B.S., Posamentier, H., Van Wagoner, J., Ross, C.A., Kendall, C.G.St.C. (Eds.), Sea Level Changes,, vol. 42. pp. 71–108.

Hardebeck, J., Anderson, D.L., 1996. Eustasy as a test of a Cretaceous superplume hypothesis. Earth Planet. Sci. Lett. 137, 101–108.

Harper, B.B., Puga-Bernabeu, A., Droxler, A.W., Webster, J.M., Gischler, E., Tiwari, M., et al., 2015. Mixed carbonate–siliciclastic sedimentation along the Great Barrier Reef upper slope: a challenge to the reciprocal sedimentation model. J. Sediment. Res. 85, 1019–1036.

Hay, W.W., 2011. Can humans force a return to a 'Cretaceous' climate? Sediment. Geol. 235, 5–26.

Hays, J.D., Imbrie, J., Shackleton, N.J., 1976. Variations in the earth's orbit: pacemaker of the ice ages. Science 194, 1121–1132.

Heller, P.L., Angevine, C.L., 1985. Sea level cycles during the growth of Atlantic-type oceans. Earth Planet. Sci. Lett. 75, 417–426.

Hilgen, F.J., Hinnov, L.A., Aziz, H.A., Abels, H.A., Batenburg, S., Bosmans, J.H.C., et al., 2014. Stratigraphic Continuity and Fragmentary Sedimentation: The Success of Cyclostratigraphy as Part of Integrated Stratigraphy, vol. 404. Geological Society Special Publication. Available from: http://dx.doi.org/10.1144/SP404.12

Hillgartner, H., Strasser, A., 2003. Quantification of high-frequency sea level fluctuations in shallow-water carbonates: an example from the Berriasian-Valanginian (French Jura). Palaeogeogr. Palaeoclimatol. Palaeoecol. 200, 43–63.

Hinderer, M., 2012. From gullies to mountain belts: a review of sediment budgets at various scales. Sediment. Geol. 280, 21–59.

Hiscott, R.N., 2001. Depositional sequences controlled by high rates of sediment supply, sea level variations and growth faulting: the Quaternary Baram delta of northwestern Borneo. Int. J. Mar. Geol. Geochem. Geophys. 175, 67–102.

Hofmann, P., Ricken, W., Schwark, L., Leythaeuser, D., 2001. Geochemical signature and related climatic-oceanographic processes for Early Albian black shales: site 417D, North Atlantic Ocean. Cretaceous Res. 22, 243–257.

Holland, S.M., 2012. Sea level change and the area of shallow-marine habitat: implications for marine biodiversity. Paleobiology 38, 205–217.

Hunter, S.J., Valdes, P.J., Haywood, A.M., Markwick, P.J., 2008. Modelling Maastrichtian climate: investigating the role of geography, atmospheric CO_2 and vegetation. Clim. Past Discuss. 4, 981–1019.

Husinec, A., Jelaska, V., 2006. Relative sea level changes recorded on an isolated carbonate platform: Tithonian to Cenomanian succession, southern Croatia. J. Sediment. Res. 76, 1120–1136.

Immenhauser, A., 2005. High-rate sea level change during the Mesozoic: new approaches to an old problem. Sediment. Geol. 175, 277–296.

Immenhauser, A., Scott, R.W., 1999. Global correlation of Middle Cretaceous sea level events. Geology 27, 551–554.

Jarvis, I., Mabrouk, A., Moody, R.T.J., de Cabrera, S., 2002. Late Cretaceous (Campanian) carbon isotope events, sea level change and correlation of the Tethyan and Boreal realms. Palaeogeogr. Palaeoclimatol. Palaeoecol. 188, 215–248.

Johnson, M.E., 2006. Relationship of silurian sea-level fluctuations to oceanic episodes and events. GFF 128, 115–121.

Joo, Y.J., Sageman, B.B., 2014. Cenomanian to Campanian carbon isotope chemostratigraphy from the Western Interior Basin, U.S.A.. J. Sediment. Res. 84, 529–542.

Jowett, D.M.S., Schröder-Adams, C.J., Leckie, D., 2007. Sequences in the Sikanni Formation in the frontier Liard Basin of northwestern Canada: evidence for high frequency Late Albian relative sea level changes. Cretaceous Res. 28, 665–695.

Kampschulte, A., Bruckschen, P., Strauss, H., 2001. The sulphur isotope composition of trace sulphates in Carboniferous brachiopods: implications for coeval seawater correlation with other geochemical cycles and isotope stratigraphy. Chem. Geol. 175, 149–173.

Kemp, A.C., Horton, B.P., Culver, S.J., Corbett, D.R., van de Plassche, O., Gehrels, W.R., et al., 2009. Timing and magnitude of recent accelerated sea-level rise (North Carolina, United States). Geology 37, 1035–1038.

Knaust, D., Bromley, R., 2012. Trace Fossils as Indicators of Sedimentary Environments, Developments in Sedimentology, vol. 64. Elsevier, Amsterdam.

Koch, J.T., Brenner, R.L., 2009. Evidence for glacioeustatic control of large, rapid sea level fluctuations during the Albian-Cenomanian: Dakota formation, eastern margin of western interior seaway, USA. Cretaceous Res. 30, 411–423.

Korenaga, J., 2007. Eustasy, supercontinental insulation, and the temporal variability of terrestrial heat flux. Earth Planet. Sci. Lett. 257, 350–358.

Krawinkel, H., Hartmut, S., 1996. Sedimentologic, palaeoecologic, taphonomic and ichnologic criteria for high-resolution sequence analysis: a practical guide for the identification and interpretation of discontinuities in shelf deposits. Sediment. Geol. 102, 79–110.

Kuhnt, W., Nederbragt, A., Leine, L., 1997. Cyclicity of Cenomanian—Turonian organic-carbon rich sediments in the Tarfaya Atlantic Coastal Basin (Morocco). Cretaceous Res. 18, 587–601.

Kuhnt, W., Holbourn, A., Gale, A., Chellai, El.H., Kennedy, W.J., 2009. Cenomanian sequence stratigraphy and sea level fluctuations in the Tarfaya Basin (SW Morocco). Geol. Soc. Am. Bull. 121, 1695–1710.

Kulpecz, A.A., Miller, K.G., Sugarman, P.J., Browning, J.V., 2008. Response of Late Cretaceous migrating deltaic facies systems to sea level, tectonics, and sediment supply changes, New Jersey Coastal Plain, U.S.A.. J. Sediment. Res. 78, 112–129.

Kulpecz, A.A., Miller, K.G., Browning, J.V., Edwards, L.E., Powars, D.S., McLaughlin Jr., P. P., et al., 2009. Postimpact deposition in the Chesapeake Bay impact structure: variations in eustasy, compaction, sediment supply, and passive-aggressive tectonism. In: Gohn, G.S., Koeberl, C., Miller, K.G., Reimold, W.U. (Eds.), The ICDP-USGS Deep Drilling Project in the Chesapeake Bay Impact Structure: Results from the Eyreville Core Holes: Geological Society of America Special Paper, vol. 458. pp. 813–839.

Laferriere, A.P., Hattin, D.E., Archer, A.W., 1987. Effects of climate, tectonics, and sea level changes on rhythmic bedding patterns in the Niobrara Formation (Upper Cretaceous), U.S. Western Interior. Geology 15, 233–236.

Laurin, J., Meyers, S.R., Sageman, B.B., Waltham, D., 2005. Phase-lagged amplitude modulation of hemipelagic cycles: a potential tool for recognition and analysis of sea level change. Geology 33, 569–572.

Leckie, D.A., Schröder-Adams, C.J., Bloch, J., 2000. The effect of paleotopography on the Late Albian and Cenomanian sea level record of the Canadian Cretaceous interior seaway. Geol. Soc. Am. Bull. 112, 1179–1198.

Li, L., Keller, G., Adatte, T., Stinnesbeck, W., 2000. Late Cretaceous sea-level changes in Tunisia: a multi-disciplinary approach. J. Geol. Soc. 157, 447–458.

Li, Q., McArthur, J.M., Atkinson, T.C., 2012. Lower Jurassic belemnites as indicators of palaeotemperature. Palaeogeogr. Palaeoclimatol. Palaeoecol. 315&316, 38–45.

Lowry, D.P., Poulsen, C.J., Horton, D.E., Torsvik, T.H., Pollard, D., 2014. Thresholds for Paleozoic ice sheet initiation. Geology 42, 627–630.

Luening, S., Marzouk, A.M., Morsi, A.M., Kuss, J., 1998. Sequence stratigraphy of the Upper Cretaceous of central-east Sinai, Egypt. Cretaceous Res. 19, 153–196.

Mart-Chivelet, J., 2003. Quantitative analysis of accommodation patterns in carbonate platforms: an example from the Mid-Cretaceous of SE Spain. Palaeogeogr. Palaeoclimatol. Palaeoecol. 200, 83–105.

Martinson, V.S., Heller, P.L., Frerichs, W.E., 1998. Distinguishing Middle Late Cretaceous tectonic events from regional sea level change using foraminiferal data from the U.S. Western Interior. Geol. Soc. Am. Bull. 110, 259–268.

Marzocchi, W., Mulargia, F., Paruolo, P., 1992. The correlation of geomagnetic reversals and mean sea level in the last 150 m.y. Earth Planet. Sci. Lett. 111, 383–393.

Masse, J.P., Fenerci, M., Pernarcic, E., 2003. Palaeobathymetric reconstruction of peritidal carbonates Late Barremian, Urgonian, sequences of Provence (SE France). Palaeogeogr. Palaeoclimatol. Palaeoecol. 200, 65–81.

McGhee Jr., G.R., 1992. Evolutionary biology of the Devonian brachiopoda of New York state: no correlation with rate of change of sea level? Lethaia 25, 165–172.

Melinte, M.C., 2004. Calcareous Nannoplankton, a tool to assign environmental changes. In: Proceedings of Euro-EcoGeoCentre-Romania, GEO-ECO-MARINA 9-10/2003-2004, pp. 1–9.

Meyers, P.A., Simoneit, B.R.T., 1989. Global comparisons of organic matter in sediments across the Cretaceous/Tertiary boundary. Adv. Org. Geochem. 16, 641–648.

Miall, A.D., 1991. Stratigraphic sequences and their chronostratigraphic correlation. J. Sediment. Petrol. 61, 497–505.

Miall, A.D., 2009. *Correlation of Sequences and the Global Eustasy Paradigm: A Review of Current Data.* CSPG CSEG CWLS convention, Frontiers + Innovation. Alberta, Canada, pp. 123–126.

Miall, A.D., Miall, C.E., 2001. Sequence stratigraphy as a scientific enterprise: the evolution and persistence of conflicting paradigms. Earth Sci. Rev. 54, 321–348.

Middleberg, J., Calvert, S., Karlin, R., 1991. Organic-rich transitional facies in silled basins: response to sea level change. Geology 19, 679–682.

Miller, K.G., Sugarman, P.J., Browning, J.V., Kominz, M.A., Hernández, J.C., Olsson, R.K., et al., 2003. Late Cretaceous chronology of large, rapid sea level changes: glacioeustasy during the greenhouse world. Geology 31, 585–588.

Miller, K.G., Sugarman, P.J., Browning, J.V., Kominz, M.A., Olsson, R.K., Eigenson, M.D., et al., 2004. Upper Cretaceous sequences and sea level history, New Jersey Coastal Plain. Geol. Soc. Am. Bull. 116, 368–393.

Miller, K.G., Kominz, M.A., Browning, J.V., Wright, J.D., Mountain, G.S., Katz, M.E., et al., 2005. The Phanerozoic record of global sea level change. Science 310, 1293–1298.

Mitchum Jr., R.M., Van Wagoner, J.C., 1991. High-frequency sequences and their stacking patterns: sequence-stratigraphic evidence of high-frequency eustatic cycles. Sediment. Geol. 70, 131–160.

Montero-Serrano, J.C., Palarea-Albaladejo, J., Martín-Fernández, J.A., Martínez-Santana, M., Gutiérrez-Martín, J.V., 2010. Sedimentary chemofacies characterization by means of multivariate analysis. Sediment. Geol. 228, 218–228.

Moriya, K., Wilson, P.A., Friedrich, O., Erbacher, J., Kawahata, H., 2007. Testing for ice sheets during the Mid-Cretaceous greenhouse using glassy foraminiferal calcite from the Mid-Cenomanian tropics on Demerara Rise. Geology 35, 615–618.

Mörner, N., 1981. Revolution in Cretaceous sea level analysis. Geology 9, 344–346.

Munnecke, A., Calner, M., Harper, D.A.T., Servais, T., 2010. Ordovician and Silurian sea–water chemistry, sea level, and climate: a synopsis. Palaeogeogr. Palaeoclimatol. Palaeoecol. 296, 389–413.

Mutterlose, J., Bodin, S., Fähnrich, L., 2014. Strontium-isotope stratigraphy of the early Cretaceous (Valanginian-Barremian): implications for Boreal-Tethys correlation and paleoclimate. Cretaceous Res. 50, 252–263.

Nelson, C.S., Hendry, C.H., Jarrett, G.R., 1985. Near-synchroneity of New Zealand Alpine glaciations during the past 750 Kyr. Nature 318, 361–363.

Norris, M.S., Hallam, A., 1995. Facies variations across the Middle-Upper Jurassic boundary in Western Europe and the relationship to sea level changes. Palaeogeogr. Palaeoclimatol. Palaeoecol. 116, 189–245.

Oppo, D.W., Fairbanks, R.G., 1989. Carbon isotope composition of tropical surface wter during the past 22,000 years. Paleoceanography 4, 333–351.

Oppo, D.W., Fairbanks, R.G., Gordon, A.L., 1990. Late Pleistocene southern ocean $\delta^{13}C$ variability. Paleoceanography 5, 43–54.

Paranjape, A.R., Kulkarni, K.G., Kale, A.S., 2014. Sea level changes in the Upper Aptian-Lower/Middle(?) Turonian sequence of Cauvery Basin, India - an ichnological perspective. Cretaceous Res. 54, 1–14.

Park, J., Oglesby, R.J., 1991. Milankovitch rhythms in the Cretaceous: a GCM modelling study. Palaeogeogr. Palaeoclimatol. Palaeoecol. 90, 329–355.

Pasley, M., Riley, G., Nummedal, D., 1993. Sequence stratigraphic significance or organic matter variations: example from the Upper Cretaceous mancos shale of the San Juan basin, New Mexico. In: Kabz, B., Pratt, L. (Eds.), Source Rocks in a Sequence Stratigraphic Framework, vol. 37. American Association of Petroleum Geologists; Studies in Geology, pp. 221–241.

Pellenard, P., Tramoy, R., Pucéat, E., Huret, E., Martinez, M., Bruneau, L., et al., 2014. Carbon cycle and sea-water palaeotemperature evolution at the Middle-Late Jurassic transition, eastern Paris Basin (France). Mar. Pet. Geol. 53, 30–43.

Phelps, R.M., Kerans, C., Da-Gama, R.O.B.P., Jeremiah, J., Hull, D., Loucks, R.G., 2015. Response and recovery of the Comanche carbonate platform surrounding multiple Cretaceous oceanic anoxic events, northern Gulf of Mexico. Cretaceous Res. 54, 117–144.

Pirazzoli, P.A., 1993. Global sea level changes and their measurement. Glob. Planet. Change 8, 135–148.

Pitman III, W.C., 1978. Relationship between eustacy and stratigraphic sequences of passive margins. Geol. Soc. Am. Bull. 89, 1389–1403.

Pugh, A.T., Schröder-Adams, C.J., Carter, E.S., Herrle, J.O., Galloway, J., Haggart, J.W., et al., 2014. Cenomanian to Santonian radiolarian biostratigraphy, carbon isotope stratigraphy and paleoenvironments of the Sverdrup Basin, Ellef Ringnes Island, Nunavut, Canada. Palaeogeogr. Palaeoclimatol. Palaeoecol. 413, 101–122.

Qie, W., Liu, L., Chen, J., Wang, X., Mii, H., Zhang, X., et al., 2015. Local overprints on the global carbonate $\delta^{13}C$ signal in Devonian–Carboniferous boundary successions of South China. Palaeogeogr. Palaeoclimatol. Palaeoecol. 418, 290–303.

Quilty, P.G., 1980. Sedimentation cycles in the Cretaceous and Cenozoic of Western Australia. Tectonophysics 63, 349–366.

Raju, D.S.N., Ravindran, C.N., 1990. Cretaceous sea level changes and transgressive/regressive phases in India—a review. In: Proceedings on Cretaceous event Stratigraphy and the Correlation of Indian Non-Marine Strata, Contributions to Seminar cum Workshop on IGCP-216, Chandigarh, pp. 38–46.

Raju, D.S.N., Ravindran, C.N., Kalyansundar, R., 1993. Cretaceous cycles of sea level changes in Cauvery basin, India—a first revision. Oil Nat. Gas Corp. Bull. 30, 101–113.

Ramkumar, M., 2008. Cyclic fine-grained deposits with polymict boulders in Olaipadi member of the Dalmiapuram Formation, Cauvery Basin, south India: Plausible causes and sedimentation model. ICFAI J. Earth Sci. 2, 7–27.

Ramkumar, M., 2015. Discrimination of tectonic dynamism, quiescence and third order relative sea level cycles of the Cauvery Basin, South India. Ann. Geol. Penin. Balk. (in press).

Ramkumar, M., Stüben, D., Berner, Z., 2004. Lithostratigraphy, depositional history and sea level changes of the Cauvery Basin, South India. Annales Geologiques de la Peninsule Balkanique 65, 1–27.

Ramkumar, M., Subramanian, V., Stüben, D., 2005a. Deltaic sedimentation during Cretaceous Period in the Northern Cauvery Basin, South India: facies architecture, depositional history and sequence stratigraphy. J. Geol. Soc. India 66, 81–94.

Ramkumar, M., Harting, M., Stüben, D., 2005b. Barium anomaly preceding K/T boundary: plausible causes and implications on end Cretaceous events of K/T sections in Cauvery basin (India), Israel, NE-Mexico and Guatemala. Int. J. Earth Sci. 94, 475–489.

Ramkumar, M., Stüben, D., Berner, Z., Schneider, J., 2010. $^{87}Sr/^{86}Sr$ anomalies in Late Cretaceous-Early Tertiary strata of the Cauvery Basin, south India: constraints on nature and rate of environmental changes across K-T boundary. J. Earth Syst. Sci. 119, 1–17.

Ramkumar, M., Stüben, D., Berner, Z., 2011. Barremian–Danian chemostratigraphic sequences of the Cauvery Basin, India: implications on scales of stratigraphic correlation. Gondwana Res. 19, 291–309.

Raymo, E., Oppo, D.W., Curry, W., 1997. The Mid-Pleistocene climate transition: a deep sea carbon isotopic perspective. Paleoceanography 12, 546–559.

Reuter, M., Piller, W.E., Brandano, M., Harzhauser, M., 2013. Correlating Mediterranean shallow water deposits with global Oligocene–Miocene stratigraphy and oceanic events. Glob. Planet. Change 111, 226–236.

Rossi, V., Horton, B.P., 2009. The application of a subtidal foraminifera-based transfer function to reconstruct Holocene paleobathymetry of the Po Delta, Northern Adriatic Sea. J. Foraminiferal Res. 39, 180–190.

Rossi, V., Horton, B.P., Corbett, D.R., Leorri, E., Perez-Belmonte, L., Douglas, B.C., 2011. The application of foraminifera to reconstruct the rate of 20th century sea level rise, Morbihan Golfe, Brittany, France. Quat. Res. 75, 24–35.

Ruban, D.A., 2015. Mesozoic long-term eustatic cycles and their uncertain hierarchy. Geosci. Front. 6, 503–511.

Ruban, D.A., Conrad, C.P., van Loon, A.J., 2010. The challenge of reconstructing the Phanerozoic sea level and the Pacific Basin tectonics. Geologos 16, 235–243.

Schwarzacher, W., 2000. Repetitions and cycles in stratigraphy. Earth Sci. Rev. 50, 51–75.

Sleep, N.H., 1976. Platform subsidence mechanisms and "eustatic" sea level changes. Tectonophysics 36, 45–56.

Spalletti, L.A., Poire, D.G., Schwarz, E., Veiga, G.D., 2001. Sedimentologic and sequence stratigraphic model of a Neocomian marine carbonate-siliciclastic ramp: Neuquen basin, Argentina. J. South Am. Earth Sci. 14, 609–624.

Spasojevic, S., Gurnis, M., 2012. Sea level and vertical motion of continents from dynamic earth models since the Late Cretaceous. Am. Assoc. Pet. Geol. Bull. 96, 2037–2064.

Strasser, A., Pittet, B., Hillgärtner, H., Pasquier, J., 1999. Depositional sequences in shallow carbonate-dominated sedimentary systems: concepts for a high-resolution analysis. Sediment. Geol. 128, 201–221.

Strauss, H., 1997. The isotopic composition of sedimentary sulphur through time. Palaeogeogr. Palaeoclimatol. Palaeoecol. 132, 97–118.

Stüben, D., Kramar, U., Berner, Z., Stinnesbeck, W., Keller, G., Adatte, T., 2002. Trace elements, stable isotopes, and clay mineralogy of thc Ellcs II K-T boundary section in Tunisia: indications for sea level fluctuations and primary productivity. Palaeogeogr. Palaeoclimatol. Palaeoecol. 178, 321–345.

Summerhayes, C.P., 1986. Sea level curves based on seismic stratigraphy: their chronostratigraphic significance. Palaeogeogr. Palaeoclimatol. Palaeoecol. 57, 27–41.

Tibert, N.E., Leckie, R.M., 2004. High-resolution estuarine sea level cycles from the Late Cretaceous: amplitude constraints using agglutinated foraminifera. J. Foraminiferal Res. 34, 130–143.

Tu, T.T.N., Bocherens, H., Mariotti, A., 1999. Ecological distribution of Cenomanian terrestrial plants based on $\delta^{13}C/\delta^{12}C$ ratios. Palaeogeogr. Palaeoclimatol. Palaeoecol. 145, 79–93.

Tucker, M.E., Gallagher, J., Leng, M.J., 2009. Are beds in shelf carbonates millennial-scale cycles? An example from the Mid-Carboniferous of northern England. Sediment. Geol. 214, 19–34.

Uličný, D., Jarvis, I., Gröcke, D.R., Čech, S., Laurin, J., Olde, K., et al., 2014. A high-resolution carbon-isotope record of the Turonian stage correlated to a siliciclastic basin fill: implications for Mid-Cretaceous sea level change. Palaeogeogr. Palaeoclimatol. Palaeoecol. 405, 42–58.

Vail, P.R., Mitchum, R.M., Thompson, S., 1977. Seismic stratigraphy and global changes of sea level. Part 4. Global cycles of relative changes of sea level. In: Payton, C.E. (Ed.), Seismic Stratigraphy—Applications to Hydrocarbon Exploration, vol. 26. Memoirs of American Association of Petroleum Geologists, pp. 83–97.

Van der Kolk, D.A., Flaig, P.P., Hasiotis, S.T., 2015. Paleoenvironmental reconstruction of a Late Cretaceous, muddy, river-dominated Polar deltaic system: Schrader Bluff–Prince Creek Formation transition, Shivugak Bluffs, North slope of Alaska, U.S.A.. J. Sediment. Res. 85, 903–936.

Van der Meer, D.G., Zeebe, R.E., van Hinsbergen, D.J.J., Sluijs, A., Spakman, W., Torswik, T.H., 2014. Plate tectonic controls on atmospheric CO_2 levels since Triassic. <www.pnas.org/cgi/doi/10.1073/pnas.1315657111>.

van Loon, A.J., 2000. The stolen sequence. Earth Sci. Rev. 52, 237–244.

Van Sickel, W.A., Kominz, M.A., Millerw, K.G., Browning, J.V., 2004. Late Cretaceous and Cenozoic sea level estimates: backstripping analysis of borehole data, onshore New Jersey. Basin Res. 16, 451–465.

Veizer, J., 1985. Carbonates and ancient oceans: isotopic and chemical record on timescales of 10^7–10^9 years. In: Sunquist, E.T., Broecker, W.S. (Eds.), The Carbon Cycle and Atmospheric CO_2: Natural Variations—Archaen to Present, vol. 32. American Geophysical Union Monograph, pp. 595–601.

Veizer, J., Bruckschen, P., Pawellek, F., 1997. Oxygen isotope evolution of Phanerozoic seawater. Palaeogeogr. Palaeoclimatol. Palaeoecol. 132, 159–172.

Veizer, J., Ala, D., Azmy, K., 1999. $^{87}Sr/^{86}Sr$, $\delta^{13}C$ and $\delta^{18}O$ evolution of Phanerozoic seawater. Chem. Geol. 161, 59–88.

Veizer, J., Godderis, Y., Francois, L.M., 2000. Evidence for decoupling of atmospheric CO_2 and global climate during the Phanerozoic eon. Nature 408, 698–701.

Voigt, S., Gale, A.S., Voigt, T., 2006. Sea level change, carbon cycling and palaeoclimate during the Late Cenomanian of northwest Europe; an integrated palaeoenvironmental analysis. Cretaceous Res. 27, 836–858.

Wagreich, M., Lein, R., Sames, B., 2014. Eustasy, its controlling factors, and the limno-eustatic hypothesis—concepts inspired by Eduard Suess. Aust. J. Earth Sci. 107, 115–131.

Wallmann, K., 2001. Controls on the Cretaceous and Cenozoic evolution of seawater composition, atmospheric CO_2 and climate. Geochim. Cosmochim. Acta 65, 3005–3025.

Wendler, I., 2013. A critical evaluation of carbon isotope stratigraphy and biostratigraphic implications for Late Cretaceous global correlation. Earth Sci. Rev. 126, 116–146.

Wendler, J., Grafe, K., Willems, H., 2002. Reconstruction of Mid-Cenomanian orbitally forced palaeoenvironmental changes based on calcareous dinoflagellate cysts. Palaeogeogr. Palaeoclimatol. Palaeoecol. 179, 19–41.

White, T., Arthur, M.A., 2006. Organic carbon production and preservation in response to sea level changes in the Turonian Carlile Formation, U.S. Western Interior Basin. Palaeogeogr. Palaeoclimatol. Palaeoecol. 235, 223–244.

Williams, D.F., Thunell, R.C., Tappa, E., 1988. Chronology of the Pleistocene oxygen isotope record 0–1.88 MY BP. Palaeogeogr. Palaeoclimatol. Palaeoecol. 64, 221–240.

Wilmsen, M., Niebuhr, B., Chellouche, P., Pürner, T., Kling, M., 2010. Facies pattern and sea level dynamics of the Early Late Cretaceous transgression: a case study from the Lower Danubian Cretaceous Group (Bavaria, southern Germany). Facies 56, 483–507.

Wyatt, A.R., 1986. Post-Triassic continental hypsometry and sea level. J. Geol. Soc. 143, 907–910.

Zhang, P.Z., Molnar, P., Downs, W.R., 2001. Increased sedimentation rates and grain sizes 2–4 Myr ago due to the influence of climate change on erosion rates. Nature 410, 891–897.

Zhao, M., Zheng, Y., 2015. The intensity of chemical weathering: geochemical constraints from marine detrital sediments of Triassic age in South China. Chem. Geol. 391, 111–122.

Zorina, S.O., 2014. Eustatic, tectonic, and climatic signatures in the Lower Cretaceous siliciclastic succession on the Eastern Russian Platform. Palaeogeogr. Palaeoclimatol. Palaeoecol. 412, 91–98.

Edwards Brothers Malloy
Thorofare, NJ USA
November 16, 2015